Food Systems and Agrarian Change

Edited by Frederick H. Buttel, Billie R. DeWalt,
and Per Pinstrup-Anderson

Searching for Rural Development:
Labor Migration and Employment in Rural Mexico
by Merilee S. Grindle

SEARCHING FOR RURAL DEVELOPMENT

Labor Migration and Employment in Mexico

Merilee S. Grindle

Cornell University Press

ITHACA AND LONDON

First published 1988 by Cornell University Press.

International Standard Book Number 0-8014-2109-8
Library of Congress Catalog Card Number 87-47970
Printed in the United States of America
Librarians: Library of Congress cataloging information
appears on the last page of the book.

The paper in this book is acid-free and meets the guidelines for
permanence and durability of the Committee on Production Guidelines
for Book Longevity of the Council on Library Resources.

To Jim, Doug, and Ward

Contents

Tables

Acknowledgments

A book such as this is always a collaborative venture. Enriched by previous scholarship, challenged by current investigation, and assisted by able colleagues, I have ample reason to acknowledge the important contributions of others to this work.

I would not have been able to undertake the study without generous financial support from the Lincoln Institute of Land Policy. At the Lincoln Institute, Sein Lin was a particularly helpful overseer of the project. John D. Montgomery of Harvard University was instrumental in ushering the project through its early stages and continued to demonstrate interest in it. Ambassador William H. Sullivan, a knowledgeable observer of Mexican development, provided consistently insightful critiques of the work in progress and was helpful in suggesting numerous revisions. Robert Paarlberg of Wellesley College was extremely helpful in providing cogent and constructive criticism of draft chapters. Judith Tendler of the Massachusetts Institute of Technology contributed greatly through her stimulating questioning of conventional understandings of economic and political phenomena. Cathryn Thorup, director of the United States–Mexico Project at the Overseas Development Council, ably coordinated and contributed to two seminars in Washington, D.C., at which the work was discussed in detail. Bruce Johnston, of Stanford University's Food Research Institute, made valuable suggestions for revisions. Colleagues at the Harvard Institute for International Development made available an environment ideally suited to a study such as this one, and HIID's director, Dwight Perkins, was gracious in his consistent support of it. Peter Timmer, Christine Jones, and Richard

Goldman of HIID added insights from a broad range of experience in rural and agricultural development in the Third World. Also at HIID, Elizabeth Shaw, Joanne Giovino, Andrea Yelle, and Cathy Collins assisted ably in many revisions of the manuscript. The support and interest of these individuals and institutions have been crucial for the completion of this book. To the extent that my work proves useful to others concerned about Mexico's future, much of the credit is due to them for their contributions.

I owe a special note of thanks to Elizabeth Meier of the Escuela Nacional de Antropología e Historia and M. Celina Robledo of the Centro de Ecodesarrollo in Mexico, who carried out much of the field work described in Chapter 4. These able researchers worked under difficult conditions to provide accurate data and human insight into the problems of four rural areas. They produced reports of great value to the overall investigation and proved to be resourceful and committed analysts of contemporary Mexican political and economic reality. Their contributions are central to this book, and I will remember our collaborative effort with pleasure for many years.

MSG

Cambridge, Massachusetts

Searching for Rural Development

I

Introduction: In Quest
of Rural Development

Throughout the Third World, rural peoples are working in locations far distant from the villages and small plots of land they continue to call home. Rural Pakistanis and Bangladeshis populate construction sites in Saudi Arabia, Kuwait, and Dubai. Indonesians from villages in West Java travel regularly to Jakarta, where they find temporary work as street hawkers and drivers of pedal rickshas. In Malawi, migrants from overpopulated rural areas move seasonally to jobs in the tobacco estates in the central region of the country. For part of every year, some 300,000 Guatemalans from highland villages migrate to other regions to work in harvesting coffee, cotton, and sugarcane. Hundreds of thousands of Brazilians move back and forth between the parched landscape of their native Northeast and the rich commercial agricultural zones in the south and west of the country. Kenyans from upland villages sojourn in Nairobi and then return to their homelands with cash to invest in the land. From Mali and Burkina Faso they travel to the Ivory Coast and the Gambia in search of temporary employment. Similar dynamics are at work when rural people in countries such as India, Peru, the Philippines, and Mexico seek jobs in distant rural areas, in urban zones, or across international borders in order to sustain their families.

The phenomenon of temporary labor migration raises important questions about opportunities for rural development in much of the Third World. For millions of households, temporary labor migration has become a normal part of living on the land. For millions more, it is a way of maintaining a rural life when no land is available to them. Increasingly, labor migration is less an option for both the landed and the landless than a necessity for their continued survival.[1] Although labor migration provides rural households with needed income, its magnitude is also a clear indictment of the limited economic opportunities available to low-income families in their home communities and a gauge of increasing dependence of rural areas on transfers of resources from elsewhere. In this sense, temporary labor migration is a symptom of rural stagnation and increasing dependence. It draws attention to the fact that many rural areas are increasingly sustained, not by agricultural activities, but by remittances of migratory income.

This pattern of rural underdevelopment and dependence has occurred most markedly in areas where the resource base for agricultural development is poor—the arid, semiarid, and rainfall-dependent regions of Africa, the Middle East, South Asia, and Latin America—and where political and economic structures impede broadly equitable access to land, capital, markets, and publicly provided resources—particularly in Latin America, the Middle East, and parts of South Asia and Africa. In such areas high levels of temporary labor migration are most apparent, and the potential of urban areas to absorb increasing numbers of permanent migrants has become most constricted. Can these regions begin to provide more secure livelihoods for the millions who are now unable to sustain themselves and their families in local communities? This book attempts to answer this question.

In recent years, development specialists have argued that technological innovations appropriate for smallholder production and institutional reforms to increase access to land and other resources generate an environment for dynamic and equitable rural development. In addition, they have argued that such changes enable the agricultural sector to play an effective role in national development by stimulating rural demand for food and manufactured goods and creating sources of productive employment in agricultural and nonagricultural activities. In particular, Bruce Johnston and John Mellor have made a case for a broad-based agricultural and employment strategy for economic development. In an article reviewing their perspective, for example, they write:

> There is a growing consensus among economists and practitioners of development who believe that a high level equilibrium of food production and employment is not only desirable on social welfare grounds but also represents a strategy capable of achieving overall growth. . . . Because of

1. For perspectives on the phenomenon of temporary labor migration in a number of countries, see Hugo (1977); Livingstone (1986b); Owen (1985); Berg (1985); Murillo Castaño (1984); *International Migration Review* (1979); Hart (1982).

the dominant position of agriculture as a source of income and employment in low-income countries, we emphasize the central importance of a broad-based "unimodal" pattern of agricultural development, characterized by gradual but widespread increases in productivity by small farmers adopting innovations appropriate to their labor-abundant, capital-scarce factor proportions. . . . reduction of malnutrition and related manifestations of poverty requires a set of interacting forces . . . that link nutritional need, generation of effective demand for food on the part of the poor, increased employment, a strategy for development that structures demand toward goods and services which have a high employment content, production of wage goods, and an emphasis on growth in agriculture. [Mellor and Johnston 1984:533]

The logic of the strategy is clear and persuasive. Indeed, the most notable success stories of rural development in the Third World came about through the effective pursuit of this strategy; countries such as Taiwan, Korea, and Japan have now become models of agriculture-based development to be emulated by other countries. Using the example of Mexico, however, the present book challenges this agriculture-based strategy of rural development. The challenge is made on the basis not of the model's theoretical importance—which is undeniably significant—but of its replicability. I suggest that a central dilemma for rural development in many countries is reconciling that which is desirable with that which is feasible politically and economically. In the concrete case of Mexico and a number of other countries, those concerned about the future of rural areas must ask a set of difficult questions.

Is rural development based on a dynamic agricultural sector feasible:

if the resource base for agriculture is poor and costly to improve?

if governments are unable or unwilling to devote significant economic and human resources to resolving problems of underproductivity and welfare in the countryside?

if land reform is not likely to occur because long-standing development policies have created a pattern of agricultural development in which a small sector of large, modern, and capital intensive farms produces important goods, generates important sources of foreign exchange, and has important access to political decision makers?

if smallholder agriculture has been increasingly abandoned in favor of temporary labor migration?

if population pressure on the land cannot realistically be relieved by extending the agricultural frontier or by encouraging permanent rural-to-urban migration?

In Mexico, where these ifs are a reality, agricultural development may not have the capacity to serve as an effective "engine" for rural development in the sense envisioned by Johnston, Mellor, and others. By exploring this issue, the book considers the potential for linking rural communities more effectively to regional and urban activities so that the problem of employment is addressed even where smallholder agricultural modernization is not a viable option. Although the following chapters focus in detail on Mexico's particular dilemma, the issues considered are broadly relevant to similar problems of rural development and employment faced by a large number of countries in the Third World.

Employment and Rural Development

Rural populations throughout the Third World are increasing in size despite the persistence of the extremely rapid urbanization of recent decades that is fueled in part by rural-to-urban migration. As indicated in Table 1, the proportion of people living in urban areas in developing countries grew significantly between 1965 and 1984. Nevertheless, the rural population increased in absolute size, a trend that is likely to continue into the future. It is estimated that the population in rural areas and in the agricultural work force may continue to grow for fifty to one

Table 1. Rural and urban growth in the Third World

Group	Urban population as percentage of total population		Rural population (millions)		Percentage of labor force in agriculture		Contribution of agriculture to GDP (percent)	
	1965	1984	1965	1984	1964	1980	1965	1984
Low-income countries[a]	17	23	1,324.4	1,839.9	78	70	42	36
Lower-middle-income countries[b]	26	37	319.3	435.4	66	56	31	22
Upper-middle-income countries[c]	49	65	163.3	173.8	45	29	17	10

[a]Thirty-six low-income economies, as categorized by the World Bank.
[b]Thirty-nine lower-middle-income economies, as categorized by the World Bank.
[c]Twenty upper-middle-income countries, as categorized by the World Bank.
Sources: World Bank, *World Tables, 1983*; World Bank, *World Development Report, 1986*; Bureau of the Census, *World Population, 1983*.

hundred years into the future (Johnston and Clark 1982:47). Moreover, during the same twenty-year period between 1965 and 1985, although the percentage of the labor force in agriculture declined, it remained relatively high for the proportion of gross domestic product (GDP) generated in the sector. These figures suggest the extent to which rural development strategies for the future must heed the problem of employment if they are to be effective in responding to rural needs.

The importance of generating opportunities for productive and remunerative employment in rural areas has certainly not gone unnoticed by development specialists or governments. In fact, strategies for rural development throughout the Third World have affirmed the importance of creating expanded opportunities for employment in agricultural and nonagricultural activities. Moreover, these strategies are broadly similar in the way that they address employment issues. In theory and in practice, most solutions to the rural employment problem are based on approaches that stress the primacy of agriculture in generating growth to spur dynamic rural and national development.[2] Development specialists consider low productivity in agriculture to be the major cause of the rural poverty and inadequate food supplies that are notable characteristics of a large part of the Third World (see Eicher and Staatz 1984; Johnston and Clark 1982; Mellor 1986).[3] As a consequence, rural development strategies are focused on the need to increase production and productivity in agriculture, and they link expanded employment opportunities to the success of achieving growth in agriculture within the context of institutions that stimulate and support equity in access to land, markets, and other resources.

At the most general level of rural development theory, greater productivity is to be achieved through the introduction of technological innovations in smallholder agricultural practices, through the provision of a variety of services and improvements—credit, technical assistance, marketing facilities, infrastructure—that give low-income farmers access to appropriate technologies and markets for their crops, and through the existence of incentives to produce resulting from appropriate food price policies. In turn, increased productivity allows farmers

2. A review of the literature on the role of agriculture and rural areas in development can be found in Thiesenhusen (1987). See also Mukhoti (1985); Staatz and Eicher (1984).

3. The problem of agricultural stagnation is particularly noted in Africa. According to one observer, "Africa is an agrarian-dominated continent where at least three out of five people work in agriculture and rural off-farm activities. Moreover, since agricultural output accounts for 30–60 percent of the gross domestic product in most countries, the poor performance of the agricultural sector over the past two decades has been a major cause of poverty and economic stagnation" (Eicher 1984:454).

and even landless farm laborers to acquire more income. More farm production stimulates the generation of off-farm rural employment in upstream and downstream industries in input supply, agricultural processing, storage, and transport. As a result of these changes, rural areas will begin to generate effective demand for a plethora of consumer goods, greater inputs into agriculture, and a variety of services, many of which can be produced or provided in rural areas using surplus labor (Johnston and Clark 1982:78–79; Ghatak and Ingersent 1984; World Bank 1978). As productivity and incomes rise, urban-rural linkages are strengthened, and opportunities in transportation, construction, handicrafts, small-scale manufacturing, trade, and services are further expanded. Greater productivity in agriculture is therefore expected to generate jobs in nonfarm activities that enhance income and welfare for both the landed and the landless (see especially Mellor and Johnston 1984; Mellor 1976; Mukhoti 1985; World Bank 1978). Through this process, a stronger, more dynamic rural economy will directly enhance national development through the efficient production of food and raw materials and expanding demand for goods and services.

This model of rural development has an empirical referent in the experience of a number of countries. In the United States, for example, agricultural development stimulated urbanization and industrialization in the nineteenth century, which in turn created opportunities for farm people to move to cities and to become part of a burgeoning industrial labor force. In the hundred years prior to 1910, the agricultural labor force dropped from 79 percent of the total labor force to 31 percent (Johnston and Kilby 1975:196). Moreover, the absolute size of the farm labor force dropped precipitously in the twentieth century, from about 12 million people at the turn of the century to 3.2 million workers in 1980 (Johnston and Kilby 1975:457; Reich, Endo, and Timmer 1986: 154). Rural areas, stimulated first by developments in infrastructure and subsequently by investments in agricultural research and extension, gradually increased production and then productivity to meet new urban demands. In the fifty-year period between 1930 and the 1980s, gross output per farm increased six and a half times (Kislev and Peterson 1986:3). In addition, land consolidation and the opening of new areas to agriculture led to an increase in average farm size from 157 acres in 1929 to 440 acres in 1982; gross output of land increased 2.8 times in the same period (Kislev and Peterson 1986:17). In response to this agricultural modernization and the increased farm income it generated, rural regions in the United States generated significant numbers of jobs in upstream and downstream activities in input supply, agroprocessing, services, commerce, transportation, and manufacturing. In the mid and late twentieth century, as farms became more heavily mechanized and capital intensive, the trend was for agriculture to become only a part-time activity for farm families. Increasingly, they combined employment in off-farm activities with the management of large, technologically advanced farms. Thus by 1982, 43 percent of all farm owners worked one-third or more of each year off the farm; off-farm employment of other farm family members significantly increased the amount of non-farm income in the total (Kislev and Peterson 1986:17). In the development of rural areas in the United States, then, agricultural productivity, rising farm incomes, and nonfarm rural employment were linked by advancements in agricultural production and productivity.

In parts of East Asia, Southeast Asia, and South Asia, productive rural economies were also created though increasing productivity in small-holder agriculture. One of the most often-cited success stories of rural development occurred in Taiwan, where a politically inspired land reform resulted in a more equitable distribution of small landholdings. Significant amounts of U.S. assistance in terms of sustaining a supportive macropolicy environment, building infrastructure, creating and staffing institutions of agricultural research, funding rural credit systems, and stimulating domestic capacity in the production of green revolution inputs were important factors in the successful transformation of Taiwan's agriculture (Johnston and Kilby 1975:243; Mukhoti 1985; Puchala and Staveley 1979:122–124). Given these conditions, an abundance of water resources, rich soil, a long tradition of labor-intensive farming techniques, investment in infrastructure, and green revolution technology transformed agricultural production and significantly raised rural incomes in a matter of two decades even though the total farm population continued to grow until the 1960s. In their study of the structural transformation of Taiwanese agriculture, Johnston and Kilby (1975:243) report that farm cash receipts more than doubled in the twenty years between 1950 and 1970. Again, advances in agricultural productivity encouraged increased demand and a diversification of the rural economy; rural households were able to generate increasing portions of their income in off-farm activities. As labor was absorbed on the farm and off, wages increased, as did the equity of income distribution. Johnston and Clark report that "by the 1960s and the early 1970s, a considerable tightening of the labor supply/demand situation led to the substantial increases in returns to labor and the reduction of poverty. . . . The initial tightening of the labor supply/demand situation was mainly a consequence of the rapid expansion of employment opportunities, especially in the nonfarm sectors of the economy" (1982: 109). Increased incomes were expended on farm inputs, consumer goods, and food, resulting in improvements in generalized levels of nutrition and welfare (see Mukhoti 1985:24–25).

Similar linkages between increased productivity in agriculture, rising incomes, and growing employment opportunities occurred elsewhere. In Japan, rapid growth in agricultural productivity characterized the postwar period, and by 1980, a minority of farm households worked at farming full time. About 70 percent of their income came from off-farm activities (Reich, Endo, and Timmer 1986:166). Thus in the cases of Taiwan and Japan, as well as in South Korea, the Punjab in India and Pakistan, Thailand, Malaysia, and a number of other countries or regions, success stories such as these demonstrate the validity of a rural development strategy focused on raising the productivity of agriculture through a relatively equitable structure of land distribution, technological innovation, and an appropriate set of macroeconomic policies. In these cases, agriculture has served as a successful "engine" of growth for dynamic rural economies that generated employment opportunities and a sound foundation for national economic development.

How widely can this model be replicated? Empirical evidence suggests that it correctly identifies a series of economic linkages between agricultural productivity and increasing incomes and welfare in rural areas. Rural development specialists affirm its utility as a way to address the issues of poverty and underdevelopment in rural areas through the generation of higher rural incomes and linkages to employment in agricultural and nonagricultural sectors. Indeed, around the world, governments and international donors and lending agencies have adopted this model as a basis for strategies of rural and agricultural development, and billions of dollars have been expended in its pursuit.

Some progress toward achieving rural development objectives in the Third World has been made in recent decades. Irrigation through a variety of water management techniques has had a significant impact on raising agricultural productivity in some areas and making possible the wider diffusion of green revolution inputs such as fertilizer and improved seeds. Increasingly, agricultural research has addressed the challenge of developing techniques and inputs that are more appropriate for semiarid and arid zones (see, for example, IDB 1986). Investments in improved and expanded transportation networks have affected large parts of developing countries' rural areas. Although the needs are still great, considerable improvements have also been made in terms of rural access to education and health facilities. In some areas, rural producers are also gaining greater access to formal institutional credit. Still, it should also be acknowledged that prior investments have already developed irrigation systems, stimulated technological innovation, and introduced economic and social infrastructure in those areas that have the greatest potential for agricultural growth. Increasingly, investments in irrigation, technological innovation, and infrastructure will be focused on areas where the payoff in terms of increased productivity is less promising (Johnston and Clark 1982:107; Lele and Goldsmith 1986). Moreover, in many of the areas that have seen the most sluggish growth of smallholder agriculture, institutional barriers to land reform and macroeconomic reform have also been greatest.

Therefore, efforts to replicate the success stories of East Asia and other regions increasingly confront severe ecological, resource, and institutional constraints. Agriculture in much of the Third World is carried out in arid, semiarid, or rainfall-dependent regions that are distinct in many ways from the regions in which rural development occurred successfully in the past. Rainfall-dependent regions present special problems for agricultural development (see Johnston and Clark 1982:93–98; Lele 1984; Mukhoti 1985:88–91). The soil is often poor in nutrients and seriously eroded. Groundwater is scarce. Rainfall is not only infrequent but generally erratic. Uma Lele (1984) in referring to the problems of improving agricultural technology in Africa notes that "low rainfall, poor soils, and the highly diverse ecological conditions within individual countries make raising agricultural productivity much more difficult in many parts of sub-Saharan Africa than in Asia, with its extensive scope for small- and medium-scale irrigation and its more fertile soils. . . . unreliable and low rainfall and poor soils . . . lead to shifting cultivation and widespread nomadism in many parts of Africa" (p. 445). Often, the crops that are grown on this parched and unproductive land and the livestock tended on it do not absorb significant amounts of labor, even when appropriate technologies are introduced. Moreover, high levels of population growth exacerbate the difficulties of raising rural productivity and welfare in much of the Third World. These difficulties have contributed to stagnating food production and decreased food security in many Third World countries. Table 2 indicates the slow growth of food production in developing countries in the decade between the mid-1970s and mid-1980s and the increasingly large imports of cereals required to feed burgeoning populations.

Equally important for many areas, particularly Latin America and the Indian subcontinent, is the inequitable distribution of landholdings in which a small number of large commercial farms occupy a significant

Table 2. Food production and food imports in developing countries

Group	Average index of food production per capita (1974–76 = 100) 1982–1984	Cereal imports (thousands of metric tons) 1974	1984	Food aid in cereals (thousands of metric tons) 1974/75	1983/84
Low-income countries[a]	116	24,017	26,430	5,651	4,878
Lower-middle-income countries	104	17,128	32,838	1,624	4,685
Upper-middle-income countries	103	24,007	52,150	705	—

[a]See Table 1 for definitions of country categories.
Source: World Bank, *World Development Report, 1986.*

portion of the best arable land and the vast majority of the rural population is pushed onto the least productive land to eke out subsistence or semisubsistence incomes. Consolidation of large landownings and population growth in peasant communities has resulted in a decrease in the average size of smallholdings in many countries of Latin America and in India, Pakistan, and Bangladesh (see, for example, Grindle 1986:121).[4] Predictably, economic growth has been concentrated in the large farm sector. Dynamic rural development under such inequitable conditions is particularly difficult. According to Bruce Johnston and William Clark (1982), this pattern of "bimodal" agricultural development that encourages capital-intensive techniques on the most productive farm holdings "will slow the rate of growth of opportunities for productive employment within the agricultural sector. In addition, the pattern of rural demand for purchased inputs and consumer goods associated with a bimodal agricultural strategy provides a relatively weak stimulus to the decentralized growth of nonfarm output and employment in small- and medium-scale firms" (p. 74). Such arguments strongly support the importance of land reform in stimulating more dynamic and equitable agricultural development.[5] Again, relatively equitable access to land was a critical ingredient in the successful de-

4. In India, for example, "it is estimated that between 1953–54 and 1971–72, a 66-percent increase in the number of rural households was associated with a 2-percent increase in the cultivated area" (Johnston and Clark 1982:107).

5. The International Labour Organisation, for example, argues that "land reform provides the only effective means of employment for the large mass of the population" (Livingstone 1986a:16–17). For a discussion of bimodal and unimodal agricultural development patterns, see Johnston and Kilby (1975); Mukhoti (1985).

velopment of rural areas in Taiwan, Japan, South Korea, and the Punjab. Effective land reforms are not easily replicated, however, because they depend to a very significant degree on the distribution of political power in a society and on the extent to which political elites fear the mobilization of the countryside. Realistically speaking, in areas where land distribution is highly inequitable, land reform will occur only when political alliances or political regimes change significantly.

A series of government policies that discriminate against agriculture in general and small farmers in particular add to the obstacles involved in replicating rural development success stories. Overvalued exchange rates that distort the prices of agricultural and manufactured exports and imports, marketing boards and food price policies biased in favor of urban consumers, cheap credit policies, and inappropriate or inadequate funding of agricultural research activities are among the culprits ascribed to strategy choices adverse to rural development. In the oft-cited success stories, the policy environment is generally credited with being particularly appropriate for providing incentives to small producers. Equally important in the cases where effective rural development has yet to occur are the disincentives for efficient smallholder production created by macroeconomic and sectoral policies. The policy environment can be changed in many countries—but not easily. Impediments to the introduction of many policy reforms are political; as such, they will be altered in most cases only through lengthy processes of political negotiation, conflict resolution, or regime change.

Thus in areas where the resource base for agriculture is poor or inequitably distributed, where the rural population continues to grow, and where policy distorts the incentives for dynamic agricultural growth, it will be particularly difficult for agriculture to serve as an "engine" of growth of the rural economy. Moreover, it is increasingly clear that the option that appeared in the United States and Western Europe—to solve the rural employment problem through urbanization and industrialization—does not offer a ready solution in much of the Third World. Already, millions of rural people have left for permanent residence in the cities. Industry has not absorbed them, government services are incapable of stretching far enough to reach them all, and major portions of the population remain in rural areas. In the Third World, employment problems that cannot be resolved in the countryside will probably not be resolved in the large cities. The case of Mexico is instructive regarding the problems inhibiting agricultural growth and employment in many of the arid, semiarid, and rainfall-dependent regions throughout the Third World.

Although this country's development has encouraged zones of highly productive agriculture, much of the central plateau region is marked by high population densities, poor soils, erratic and insufficient rainfall, and a long history of periodic crop failure. Land distribution is highly inequitable—conforming to a classic pattern of large-scale capital-intensive farms and those that barely permit peasants to eke out a subsistence living—yet a significant land reform for the purpose of altering this pattern is probably not likely, given the domestic and international context of the country's development (see Chapter 3). The rural population continues to grow in size, as does the incidence of landlessness, poverty, and malnutrition. Government policies to control the price of basic food items and to channel credit, appropriate technology, technical assistance, and infrastructure to zones of highly profitable agriculture have also played a major role in creating an underproductive sector. To exacerbate the dilemma faced by Mexico's rural poor, permanent migration to the country's urban areas is no longer a reasonable option, because current levels of urbanization underscore the limited capacity to generate productive jobs and to meet basic needs for shelter, services, education, and health. Table 3 indicates that both rural and urban zones will continue to increase significantly in population for the foreseeable future. Confronting these constraints, rural inhabitants have become increasingly dependent on temporary labor migration opportunities.

A study of fifteen agricultural development projects oriented toward small farmers in Mexico reported that households in ten of the project areas were heavily involved in migratory labor in addition to farming (see CEICADAR 1984:27). Taylor (1984a:68) discovered that in a single year (1982) nearly 40 percent of his sample in two villages in Michoacán engaged in some form of temporary labor migration. Indeed, considerable evidence from a variety of local settings indicates

Table 3. Real and projected growth rates for urban and rural population, Mexico (percent)

Item	1970–75	1975–80	1980–85	1985–90	1990–95	1995–2000
Total	3.2	3.0	2.9	2.7	2.4	2.2
Urban	4.3	4.0	3.8	3.5	3.1	2.8
Rural[a]	1.5	1.2	1.0	0.8	0.6	0.3

[a]Fewer than 2,500 inhabitants.
Source: *Statistical Abstract for Latin America,* 1983, p. 89.

that the income from seasonal or temporary migration forms a major portion of overall household income. In one village in north-central Mexico, more than half of household income was derived from temporary labor migration (Mines 1981:47). Similar dependence on migration income was discovered by Cornelius (1976b) in his study of nine communities in rural Jalisco, where at least 50 percent of the economically active population had engaged in labor migration. Studies of communities in the states of Michoacán, Nuevo León, Oaxaca, and Mexico also reported that high rates of temporary labor migration were common (see Gregory 1986:109, 113). In one village in Oaxaca, more than 90 percent of households were reported to depend in some part on income from migration, and in Zacatecas a similar situation was faced by nearly 80 percent of village families (Stuart and Kearney 1981:35; Mines 1981:24). The location and security of employment opportunities are critical issues for Mexico's rural population; for the rural areas that the migrants call home, the issues relate to the potential to achieve more broad-based development.

Indeed, the underdevelopment of much of rural Mexico has not gone completely unnoticed by the government. Since the early 1970s, the government has made efforts to increase the productivity of rain-fed agriculture and to increase rural standards of welfare and to ensure national food self-sufficiency. The programs focused on rain-fed areas that have been pursued are legion; from the mid-1970s through the early 1980s, the government spent annually about 15 percent of the central budget for agricultural, forestry, and fishing development. In 1982 this sum amounted to 104,950.6 million pesos ($1,908.2 million in 1982 dollars) (de la Madrid 1983:511). A major national food policy was budgeted for nearly $7 billion between 1980 and 1982 (Grindle 1985). Additional large sums were spent on infrastructure, health, sanitation, and education aspects of rural development. In short, rural development based on improving the potential of rain-fed agriculture has a substantial history in Mexico.

The results of this history of heavy investment have been disappointing. Overall agricultural growth rates have been considerably below rates of growth of GDP for most years, as Table 4 makes evident. Agricultural production on rain-fed plots remained practically stagnant throughout the 1970s, grew considerably in the early 1980s under the dual impact of good weather and the availability of extensive subsidies to production, and then declined again in 1982 and 1983 (de la Madrid 1983). The need to import massive amounts of basic staples remained a

Table 4. Economic growth rates for Mexico, 1971–1985

Item	1971–1975	1976	1977	1978	1979
GDP growth	5.7	2.1	3.3	7.0	9.2
Agriculture	1.7	1.2	5.1	3.6	−2.1

	1980	1981	1982	1983	1984	1985
GDP growth	8.3	7.9	−0.5	−5.3	3.7	2.7
Agriculture	7.1	6.1	−0.6	2.9	2.5	2.2

Source: IDB 1979:14; 1982:281; 1986:314.

major problem for the government in the 1970s and emerged again as a problem in the mid-1980s (see Table 5). Perhaps more important, levels of rural unemployment, underemployment, and landlessness all increased despite markedly increased government investment, and these conditions stimulated unprecedented levels of labor migration from rural areas.

Mexico's recent experience with rural development strategies that center on increasing production and productivity in agriculture provides reason to be cautious about the capacity of such an approach to make income and employment opportunities more available in rural areas. As in many parts of the Third World, agriculture-led rural development in Mexico is severely constrained by the poor agricultural resource base, population pressure on the land, inequitable access to land, and a variety of discriminatory policies. Moreover, where some potential for agricultural growth exists, the costs of increasing production may be great, outstripping the value of the increased output, especially of the staple crops that most government programs are designed to encourage. Given the poor agricultural resource base, directing appropriate technology, infrastructure, irrigation, credit, and technical assistance to these zones may not be the most efficient use of limited public resources and may not offer attractive incentives to peasant farmers to invest more time and effort in their plots. At issue, then, is whether alternative "engines" for rural development can be found for areas in which raising agricultural production and productivity may not be feasible under current conditions and where the prospects for significantly altering these conditions are not encouraging.

An approach to rural development that focuses on the employment issue should more readily address the complex problems of a frequently poor resource base for agriculture, population pressure on the land,

Table 5. Trade balance of three basic staples, Mexico, 1960–1980 (thousand tons)

Year	Corn	Wheat	Beans
1960	428.966	−4.238	−2.267
1965	1,335.155	672.412	16.016
1970	−759.197	40.597	2.684
1971	256.103	−91.332	−.313
1972	221.683	−624.576	35.871
1973	1,113.595	−707.174	10.710
1974	1,280.529	−956.532	−38.753
1975	−2,654.550	−43.462	−104.214
1976	−909.635	15.703	42.022
1977	−1,984.236	−430.920	100.820
1978	−1,342.702	−437.014	42.864
1979	−744.781	−1,147.135	−5.395
1980	−4,186.643	−899.000	−440.928

Source: Spalding 1984:2.

increasing landlessness, and institutional rigidities. To develop such an approach, cues can be taken from the choices made by rural households. For most of them, the option of permanent rural-to-urban migration is rejected as no longer viable.[6] Similarly, the option of committing all their resources toward generating self-sufficiency and an agricultural surplus on small plots of poorly endowed land is increasingly rejected in favor of one that diversifies the sources of their income. Moreover, for the landless, the option of access to land and subsistence production is increasingly foreclosed. The creation of more viable rural communities, then, may depend upon the ability to respond directly to needs for productive sources of employment rather than assuming that these will be met through increases in agricultural productivity. The employment issue and its implications for rural development are reasons why this book begins with an interest in understanding the phenomenon of temporary labor migration.

Migration, Employment, and Policy Analysis

Temporary labor migration appears to be a universal response of poor rural households to the limitations of agricultural development in

6. In a study of a large region in the state of Oaxaca, it was determined that, of every ten inhabitants, three would move permanently from the rural area, four would be temporary labor migrants, and three would remain permanently in the community (Kearney 1984:25, 27).

their villages and regions. It indicates a readiness to search for employment opportunities wherever they may be found, to diversify sources of income, to generate cash that is increasingly needed even for subsistence, and to maintain the possibility for continued existence in rural areas. Rural people, stimulated by the exigencies of daily life, are searching for alternative "engines" for rural development. Interestingly, most evidence suggests that migrants are inventive and successful in finding jobs and generating income for their families. Whether through wage labor or self-employment in urban and rural informal sectors, the issue is less one of direct unemployment or even underemployment than one of the location and productivity of the jobs that are secured and the future of the rural communities themselves.

Those who leave rural communities in search of work are not hapless victims, flung out in random fashion to search for the means of survival. Instead temporary labor migration is a strategy adopted by households in conjunction with other decisions about how to use available labor most effectively and how to add to general household income and welfare, both in the short term and in the longer term. The decision to migrate—how it is made, why it is made, what adjustments it requires within the household, and what economic possibilities it creates for the household—is of central importance to policymakers searching for a better "second-best" solution to growth and employment problems in rural and semirural areas. The incentive to migrate is therefore a starting point for this book because it focuses attention on the needs and experiences of rural households and indicates the extent to which rural families have become dependent on nonlocal sources of income.

Just as rural households have become dependent on the possibility of generating income elsewhere, many local communities have become dependent on the infusion of migratory earnings. Consequently, policy analysts need to direct considerable attention toward locally and regionally specific characteristics of the labor migrants' places of origin and their potential for generating sources of productive employment in agricultural and nonagricultural pursuits. An employment-focused development strategy encourages an understanding of the important links between rural, semirural, and urban areas and suggests the importance of adopting an inclusive definition of rural areas. Rural areas should be understood to include not only small farming communities but also the market centers and towns that are linked to the farming communities through economic ties to agricultural and nonagricultural activities (see Anderson and Leiserson 1980:227–228). The creation of employment opportunities in these communities, market centers, and towns can respond to the needs of rural inhabitants and can stimulate more balanced regional economic growth. In light of the importance of this analytic issue, the following chapters focus on the central plateau region of Mexico, and the book includes a closer assessment of four specific areas in this region.

The development of these rural and semirural regions cannot be accomplished by focusing on the local communities in isolation, however. Their futures are tied to decisions made by policy elites in capital cities. The policies that these elites establish to affect trade, interest rates, exchange rates, taxes, and decentralization are critically important for rural areas. Central to the following analysis, therefore, is the question of the political, economic, and administrative feasibility of various policy options that are determined by national decisionmakers. Thus an important purpose of this book is to go beyond analysis of a particular set of problems and the discussion of a set of alternatives to a consideration of how needs and recommendations can be converted into political initiatives for reform. In most political systems, there are biases against national development strategies that focus on the needs of rural areas and rural people. Very possibly for this reason most considerations of rural development problems do not explore sources of the initiative for reform. Although the potential for reform must be assessed specifically for each country, the analysis presented in the Mexican case study should stimulate consideration of the politics of reform elsewhere.

Finally, the international context for rural development as it affects trade, prices, and opportunities for migration suggests that problems and prospects must be considered from an international perspective as well as from a domestic one. The implication is clearly that policy analysts need to understand the policies themselves and also the broadest economic and political contexts in which they are formulated and implemented. The international context of rural development has particular importance for Mexico because of its close proximity to the United States and the extensive interrelationships of these neighboring economies. This book therefore includes a discussion of significant aspects of relations between the United States and Mexico as they are relevant to the selection and pursuit of present and future rural development strategies.

Because of the importance of these four interrelated issues, the analysis in the following chapters will integrate data and research questions at both the micro level of the household and community and the macro level of national and international contexts. The chapters that follow explore the rationale for pursuing an employment-oriented rural development strategy. Chapter 2 examines extensive temporary labor migration as it illuminates the nature of economic opportunities in Mexico's rural areas and the way in which rural households respond to the opportunities that are available to them. The discussion reflects a model of household income-generating strategies that demonstrates why allocating household labor to migratory activities is a reasonable solution to the problem of ensuring family welfare, particularly during periods of economic crisis. The poor quality of land in much of Mexico's central plateau, dependence on frequently erratic rainfall for production, population pressure and increasing landlessness, discriminatory government policies, the concentration of industrial employment in a few very large cities, a need for cash income, limited opportunities for local off-farm employment—these are some of the most important reasons why labor migration has achieved importance for large numbers of rural households in Mexico. At the same time, labor migration makes it possible for poor rural households to remain in rural areas, offering them the option to resist moving permanently to already overburdened urban areas. They make additional choices in deciding who becomes involved in labor migration and in selecting destinations for the migrants who leave their communities to search for a job. Chapter 2 indicates how and why labor migration is useful to rural households but also considers the extent to which it has a negative impact on the local community; I suggest that the rural development of the country has failed to provide rural inhabitants with other options.

Chapter 3 assesses a variety of options for creating employment in Mexico's rural areas in light of what has been learned about rural development and household decisionmaking. The chapter posits that a central goal of a successful rural development strategy should be the creation of productive sources of employment and income in agricultural and nonagricultural pursuits. The activities that need to be considered include agricultural development, land reform, investments in infrastructure, rural industrialization, and the mobilization of remittance income. None of these alternatives offers easy solutions to Mexico's rural employment problems. In particular, a discussion of the potential for agricultural development in the country indicates that it cannot always be the "engine" for rural development and that it does not easily produce increased numbers of jobs. If this is the case, alternative sources of investment in rural and semirural areas need to be considered. Not only for agricultural development, then, but also for employment generation, rural areas should be more clearly linked to market towns and small cities and should be involved in regionwide industrial development. In most cases, the specific options available to local communities will vary, depending on their natural resource endowment, their location, their historical development, and their preexisting links to regional, national, and international economies.

Chapter 4 considers the differences among rural communities and the options available to them in greater detail and reports data from field research carried out in the summer and fall of 1985. Profiles of four rural areas indicate how options in each are shaped by the nature of agricultural development, local structures of power and influence, geographic location and linkages to markets, and the tradition of migration peculiar to each. They also suggest the extent to which remittance income has been responsible for keeping rural communities viable in the past and present. Through the experience of a number of households, the chapter demonstrates how migratory income has been invested in productive activities that make it possible for migrants and their families to maintain their rural base. The future of these four areas depends in large part on the extent to which they can overcome the particular constraints that make labor migration a rational choice for their inhabitants.

On the basis of the options considered in Chapters 3 and 4, Chapter 5 indicates the reforms needed to stimulate more employment-generating activities in rural areas and considers the potential to adopt and implement these changes. Characteristics of the Mexican political system are central determinants of the way in which policy reform is brought about. Centralization of power in the executive and the relatively closed nature of decisionmaking mean that officials within the government must adopt a reoriented approach to rural development as a "project" and must achieve presidential support if they are to be effective as reformers. They must also consider the extent to which the political system institutionalizes "slippage" in the allocation of public goods and services as a strategy for keeping the political peace; reformers must therefore be concerned not only about the politics of decisionmaking within the executive branch of government but also with the politics of the implementation process. Moreover, they are not likely to have strongly mobilized rural support to help them achieve their reforms; rural inhabitants have generally chosen migration and other economic strategies rather than political activism as a method of responding to the difficult circumstances they face. Chapter 5 also documents the experience of prior rural development efforts in the country and its meaning for reformers in the present and future, who will be hampered even further by a lack of resources to bring about needed changes. The task of reform will not be an easy one in Mexico because of the uncertain sources of political support and lack of resources. In all likelihood, rural development will become a national priority only when it becomes clearly linked to the extensive and troubling urban problems that Mexico faces.

The final chapter assesses the options for U.S. policies that can facilitate a more promising development strategy in Mexico. The future of Mexico's rural areas depends fundamentally on decisions that must be made in Mexico, but U.S. policies about trade, migration, and economic assistance affect the capacity to pursue a rural development strategy more focused on the employment issue. The principal concerns of the United States—to ensure stability on its southern border, to limit illegal migration from Mexico, and to enhance trade and investment opportunities—can effectively be facilitated by encouraging Mexico to develop its rural regions more equitably and rapidly. The reality of U.S.-Mexican relations, however, makes it difficult for Mexico and the United States to discuss their differences with regard to these issues. The final chapter, then, relates the problems of individuals and households in rural communities to the priorities of national and international decisionmakers.

Evidence from the Field: An Explanation

There is a rich literature on rural communities in Mexico, the contribution of anthropologists, sociologists, political scientists, economists, journalists, and others who have been at once fascinated by the rich heritage of these areas and frustrated by their poverty and underdevelopment. The present book draws extensively on this literature. In addition, it uses official documents and censuses to illuminate aspects of Mexico's rural development history. Public officials from several administrations have been interviewed and have contributed valuable insights into the issues of rural development and public policy in Mexico. The book also presents evidence accumulated during field research carried out in four rural *municípios* in the central plateau area of the country.[7] These municípios, with populations varying between thirteen thousand and twenty-two thousand inhabitants, manifest a variety of problems and prospects that affect a large number of rural communities

---

7. A município is a geopolitical unit corresponding roughly to a county in the United States. There are some two thousand municípios in Mexico.

in Mexico. Tepoztlán in the state of Morelos, Jaral del Progreso in the state of Guanajuato, Unión de San Antonio in the state of Jalisco, and Villamar in the state of Michoacán represent the present and possible future of many similar municípios. Each município is composed of a municipal center, or *cabecera*, which is the political and administrative center of the area and usually its economic focus also, and numerous villages or *ranchos*, small communities that are politically and administratively ascribed to the cabecera. The objective of the field research was to generate an economic and political profile of each of the municípios in terms of the daily life of the cabecera and two, three, or four of the villages ascribed to it.

Although a large number of Mexico's rural municípios would have been appropriate sites for the research conducted, several criteria were used in selecting Tepoztlán, Jaral del Progreso, Unión de San Antonio, and Villamar for more detailed analysis. First, areas were selected that were far enough removed from Mexico City that they could be characterized as having a local economy that was not simply an extension of the economy of the capital city. The closest município to Mexico City, and the most urbanized, is Tepoztlán, approximately eighty kilometers from the capital. Although Tepoztlán is within the economic orbit of Mexico City, it retains a distinct regional and local economy. The other municípios are about 250, 400, and 430 kilometers from Mexico City, respectively, and each is characterized by an identifiable local and regional economy.

Another important criterion for selecting the four municípios was the fact that each has been studied in the past by other social scientists. Thus existing historical information provides insight into the transformations that have occurred over time in these rural areas. This historical information is particularly important because of the severe nature of the economic crisis affecting Mexico in the 1980s; longer-term trends could be separated from more immediate responses to the crisis. A third criterion for selection was that the four municípios demonstrate a range of agricultural conditions. Thus two areas were chosen that appeared to have some potential for agriculture-based rural development—that is, areas with irrigation or adequate rainfall and productive soils—and two areas that appeared to lack much potential for the development of agriculture. This criterion was used in order to explore more fully the kinds of employment-creating activities that might be stimulated where agriculture generates little income.

The research conducted by the author and two researchers was not meant to be exhaustive in describing the four municípios. Rather, its purpose was to generate a brief descriptive overview of each in order to highlight its development in the past, the dilemmas its inhabitants face in the present, and the alternatives for the future. Past, present, and future are linked to the process through which rural households find jobs that can sustain them and at times allow them to improve their welfare. Each of the communities has had a somewhat distinct history of labor migration. In one area, for example, labor migration is a relatively recent phenomenon and is directed primarily toward industrial centers in Mexico. In another area, a strong tradition of migratory labor dates to colonial times and has been directed toward the United States since the nineteenth century. In two regions, labor migration has brought considerable evidence of modernization and development—albeit dependent—to local communities. In all cases, however, the incentive to migrate is rooted in the lack of economic opportunities in the sending communities. This book is written in the belief that understanding labor migration as a response to rural underdevelopment can lead to more effective strategies for future development.

2

Deciding to Migrate:
Pushes, Pulls, and Portfolios

Labor migration has become an integral aspect of life in rural Mexico. For hundreds of thousands of families, the search for work outside the local community has become a critical component of complex strategies for ensuring survival, for coping with unexpected economic demands, and for investing in a more secure future. Migrants search for work in other rural areas, in small towns and cities, in the already overcrowded urban centers of the country, and in the United States. The frequency with which rural inhabitants leave their communities has increased significantly in recent years as demographic and economic conditions have made it increasingly difficult to achieve a decent standard of living.

In this chapter, the causes and consequences of high levels of temporary labor migration are considered in an effort to understand the dynamics of the process and what it means for rural households and rural communities. What fuels the migratory process? Why do some people choose to migrate and others to remain in the local community? Why do some communities and regions have higher rates of migration than others? What determines where migrants will go in search of work? What happens in a household while the migrant is away? What happens in a community when migrants return? Responses to such questions provide a perspective from which to assess options for rural development in Mexico, a task undertaken in Chapter 3.

It is, of course, difficult to know with precision how many rural households in Mexico engage in temporary labor migration. National census data provide insights into shifting population trends but provide no basis for estimating the seasonal and temporary movements of individuals in search of employment opportunities. Similarly, because most policy concern at the national level is directed toward the issue of urbanization and its consequences, nationwide surveys of population movements also tend to focus on patterns of permanent migration.[1] Many of the data on temporary labor migration that are available come from intensive analysis of individual communities and regions in which high rates of rural outmigration have been noted. Fortunately, a number of relevant studies present great insights into the local context of migration, its causes, and its consequences. They are relatively recent, display generally high quality, are quite diverse in the kinds of communities examined, and are remarkably consistent in their findings regarding major questions relating to labor migration and rural development policy. Nevertheless, although such studies provide a broad and deep base of ethnographic information, they are unable to offer much guidance about national trends and experiences.

Similarly, although the causes and consequences of Mexico–United States migration have been extensively studied, there are important limitations on the use of this data source. Because much Mexico–United States migration is undocumented, most studies can only estimate its magnitude and often base their findings on severely flawed data collected by the Immigration and Naturalization Service on apprehensions of individuals attempting to enter the United States illegally (see Conroy 1981:27; Gregory 1986:172–175). Moreover, from such data it is difficult to know how many of the migrants are motivated by a search for temporary or permanent employment. Thus estimates of the annual net inflow of labor migrants from Mexico range from a high of 1 million to a more modest 180,000 individuals. The range of variation is apparent in a sampling of studies. A national survey conducted in Mexico, for example, indicates that in 1978, at any given moment, between 500,000 and 625,000 Mexican laborers were found in the United States (Zazueta and García y Griego 1982:51). The 1980 U.S. census counted 1.13 million undocumented workers from Mexico, 55 percent of the total number of such workers counted (Passel and Woodrow 1984:11–12).[2] Clark Reynolds estimates that a million temporary migrants enter the

1. National survey data on temporary migration are difficult to gather, as such migration is highly seasonal; thus the timing of data collection can easily lead to skewed results. For a discussion of data collection problems, see Teitelbaum (1985: chap. 3).

2. No other sending country accounted for more than 5 percent of the total (see Passel and Woodrow 1984).

United States each year and that the number is increasing. He suggests that 10 percent of the growth of the U.S. labor force consists of Mexican migrants. Furthermore, he indicates that temporary labor migration accounted for 5.8 percent of the Mexican labor force in the mid-1970s (Reynolds 1979:123).

Studies have indicated that at least 70 percent of the labor that eventually migrates into the United States originates in rural areas, and the majority of migrants come from villages of fewer than 2,500 people (Mines 1981:23–24). Furthermore, there is considerable consensus among researchers that the majority of temporary labor migration to the United States originates in eleven states of the central plateau area of Mexico (about 60 percent) and six northern border states (about 11 percent) (Mines 1981:16; Zazueta and García y Griego 1982:61). The central plateau tends to send mostly rural migrants to the United States; the northern border region tends to send more urban migrants, although in many cases these individuals migrated north at an earlier time from rural areas (Mines 1981:16). There is also considerable evidence that temporary domestic migration often leads to subsequent U.S. migration (see, for example, Dinerman 1982:69; Kearney 1984; Sánchez and Romo 1981:4–5). The extensiveness of labor migration to the United States is suggested in a study of communities in central Mexico in which it was discovered that 90 percent, 59 percent, and 30 percent of households in the respective communities had sent migrants to "El Norte" (Verduzco 1984:42).

These data must be understood as subject to considerable error; nevertheless, all confirm what village-level studies suggest: that labor migration is massive and that it is increasing over time. Numbers alone, however, do not allow analysts to understand the dynamics of labor migration among rural people. To interpret the data, they must use a theoretical framework that addresses issues of causality. In discussions of migration, theories tend to focus on the question of whether rural people are "pushed" or "pulled" into migratory activities.

Pushing or Pulling? Theories to Explain Labor Migration

In the past, theories of migration generally tended to focus on the nonlocal causes of the process. In an early formulation, for example, it was argued by W. Arthur Lewis (1954) and others that the process of industrial development in Third World countries created a dual economy with significant implications for employment and urbanization. According to the theory, employment opportunities in the developing modern industrial and urban sectors provide the capacity to absorb surplus labor from the traditional agricultural sector. A generalized process of pulling surplus labor into the urban areas ensues, helping to hold wages down in industry and thus spurring the drive toward industrialization (see Berry and Sabot 1984 for a discussion).

In recent years, this model has been largely abandoned by students of migration and development, for it is unable to explain two of the most visible aspects of much contemporary labor migration. The first is the disjuncture between continued high rates of rural-to-urban migration and a clearly apparent lack of industrial jobs in urban areas. In Mexico, for example, the manifest inability of a capital-intensive industrialization process to absorb the extensive supply of surplus rural labor has spawned a very large urban informal sector, a result not anticipated in the dual economy model. Second, the theory attempts to explain permanent rural-to-urban migration but has little to say about the temporary nature of much labor migration in Mexico today.

Later, Todaro's (1969) influential work stressed the importance of wage rates as an impetus to migration (see also Harris and Todaro 1970). In this perspective, potential migrants make rational choices as to where they anticipate the highest wages and are influenced by expected future gain in wage earnings. Thus individuals may decide to migrate, even when it is well known that urban or industrial jobs are scarce, in the expectation that at some future point in time they will benefit from a job with higher wages and be better off than they could expect to be in their place of origin. In the Mexican case, Todaro's model helps account for the attractiveness of the U.S. labor market (where wages are relatively high and employment opportunities available) over the domestic urban labor market (where wages are relatively low and jobs more scarce). Still, the model is not fully able to explain why some individuals choose domestic destinations in preference to others or why so much migration is temporary and circular rather than permanent (see critiques by Dinerman 1982:4–5; Kearney 1984:14; Rhoda 1983:40; see also Garrison 1982).

Moreover, Todaro's model is based on the assumption of rational economic behavior by individuals. As we will see, however, decisions to migrate are generally made on the basis of a calculus of household income streams from a variety of sources—agriculture, wage labor, petty commerce, artisan manufacturing, and remittances from labor migration—and so level of wages may not be as important as the extent to which employment opportunities combine efficiently or effectively with other income-generating pursuits (see Taylor 1984a:7–8). Thus, for example, some household members leave the local community in search of income-generating activities for a relatively short period, often for six to eight months at a time. Other household members remain in the community, cultivating land, tending livestock, producing artisanry, pursuing commercial activities, or searching for wage labor locally. The constrained rationality of households, not individuals, may therefore be the most appropriate decision unit to explain.

At the broadest level, some explanations of labor migration focus on the characteristics of capitalist development at the national and international levels; the dynamics of capitalist accumulation, they argue, require that a cheap and mobile pool of labor be available. Rural conditions are maintained or depressed to such a level that surplus labor is forced into a national and international migratory process that keeps wages and consumer prices low. Labor migration itself is encouraged to be temporary so that the rural base can help "reproduce" the reserve supply of labor. This model is important for its demonstration of the role of the state in pursuing policies to encourage both capitalist accumulation and labor surplus (see Piore 1979). Once again, although this explanation suggests reasons for the need to migrate, it is less helpful in explaining particular choices as to when to migrate and when to stay at home, where to go, and when to return and fails to indicate who migrates and who does not. From the perspective of a policy analyst, these questions are of great importance in attempting to understand the role of economic options in rural areas and the potential to expand them in ways that can provide some alternative to the need to migrate.

In contrast to theories that focus on extralocal influences in migration, most community-based research on migration has come to emphasize the local roots of the decision to migrate. Household decision models are often used as a means of exploring the why, who, how, when, and where questions about migration (see Dinerman 1982; Taylor 1984a, 1984b). They consider the question of the local resource and opportunity base available to rural households and are able to deal with the accumulated evidence about the selectivity of migration among individuals, households, communities, and regions. Such models posit that households attempt to maximize total family income at the same time that they attempt to minimize the risk to overall income level through diversification. Thus some opportunities are forgone in favor of others, and labor allocations and investment decisions are continually reassessed in terms of a diversified household "portfolio" of income-generating activities (see Roberts 1982, 1985).

The models allow for the possibility that decisions themselves are based on a rational calculus of cost and benefit—at the household, not the individual, level—and that information (about wage rates and kinship networks) is important in shaping decisions (see, for example, Stark and Levhari 1982; Taylor 1984a, 1984b; Yap 1977). Furthermore, they allow for the fact that households and the decisions they make are constrained by the development potential of their community, the policies historically adopted by governments that affect them, and the characteristics of both national and international political economies. Finally, they seem quite appropriate for explaining circular migration patterns by allowing for the possibility that the individual migrant leaves the community temporarily in order to sustain the household in the rural area and to contribute to its survival or development over time. The motivation to migrate is to be found within the rural community itself. This perspective is therefore useful in responding to the question of why migration occurs, who migrates, where migrants go, and what they do when they return to their rural communities. It is less useful, however, in assessing the potential to change the broad range of conditions that encourage migration. Thus in later chapters, discussions of national development policies, political decisionmaking, and international constraints indicate the importance of a more general political economy perspective to assess the future of rural development in Mexico.

Why Migrate? The Rural Opportunity Base

Migration and employment issues in rural Mexico reflect dramatic changes in the country's demographic profile since the 1960s. In 1980 70 million people lived in the country, as compared with 35 million people just twenty years earlier. Projections for the year 2000 indicate a total population of 109 million inhabitants, a figure that is expected to increase to 182 million by the year 2050 (World Bank 1984:193; see also Kirk 1983). Several decades of rapid population growth prior to the mid-1970s produced a high proportion of economically active people in the total population; moreover, many of those born during periods of high population growth entered the labor market for the first time in the 1970s and 1980s. In 1980 the economically active population (23.9 million persons) was more than twice as large as it had been in 1960 (11.3 million persons), and by the 1980s this labor force was growing at a rate of 3.3–4.0 percent annually (Gregory 1986:20; Alejo 1983:83–87; *LAM* 1986:30; Reubens 1979:132; World Bank 1984: 259). In concrete terms, this level of growth meant that, in 1980, approximately 800,000 individuals joined the labor force; by 1985, 1 million new entrants a year were added, and this number was expected to increase to more than 1 million annually until the year 2000 (Alejo 1983:83–87).

The pressure of these numbers on the country's labor market is striking. Although Mexico has achieved considerable progress in becoming a rapidly industrializing Third World country, the process of development itself has not created the level of industrial employment appropriate to its population base. A significant portion of the labor force has been absorbed into the informal or tertiary sector of the economy, and the size of the agricultural labor force continues to exceed its contribution to GDP (see Gregory 1986).[3] Until the 1980s, wages appear to have grown over the course of several decades, although gaps in the distribution of income widened (see Gregory 1986). Severe crises of the 1980s, however, brought a sharp contraction in economic growth rates at the same time that the labor force continued to grow rapidly. Employment became an even more significant issue of public concern when a reported 2 million jobs were lost in 1982 and 1983 because of the economic crisis (*LAM* 1986:40). By 1985 the government was reporting a 10.7 percent unemployment rate, and underemployment was estimated to affect 50 percent of the working population (*LAM* 1986:39–40). Although these figures probably understate the absorptive capacity of the informal sector, there is little question that the economic crisis of the 1980s exacerbated the tension between population growth and the creation of sources of productive employment.

The issue of employment is critical in the country's rural areas, where 25 million people—most of them poor—live; they include 40 percent of the economically active population. Many have already abandoned rural areas in search of employment in urban zones. In Mexico, at least

3. Using aggregate data, Peter Gregory argues that labor market and wage-level growth in Mexico have been considerable and that labor productivity is impressive by developing country standards. Between 1970 and 1980, he reports that the labor force in the primary sector grew by 0.2 percent annually, in the secondary sector by 5.3 percent, and in the tertiary sector by 6.5 percent. In all cases, output per worker increased annually (Gregory 1986:31). His detailed review of the evidence leads him to argue that the "informal sector" of the economy has been able to generate considerable productive employment. The contention here is not that rural inhabitants are unemployed but rather that to generate income they are forced to search far and wide (usually successfully) for jobs.

since the 1940s, there has been steady and massive outmigration from rural areas to large urban centers, particularly Mexico City, Monterrey, and Guadalajara. In 1980, 69 percent of the population lived in centers of 20,000 inhabitants or more, and 48 percent of the urban population lived in the seven largest cities in the country, 32 percent in Mexico City alone (SALA 1983:5; World Bank, 1984:261). Population growth rates for both urban and rural areas in Mexico indicate substantial urbanization, beyond the contribution of natural population increase (see Table 1). Such figures begin to suggest the extent to which rural families in Mexico have abandoned their rural roots to seek employment opportunities in urban settings (see Gregory 1986: chap. 5). Permanent migrants demonstrate a preference for larger urban areas rather than smaller regional towns and cities (see Cornelius 1976b:11). From a number of studies, we know that this population shift from rural to large urban areas is permanent; once whole families have left their rural place of origin, they seldom return for more than short visits. Moreover, with time their ties to the rural community tend to loosen (see, for example, Cornelius 1976b).

In spite of the extensive rural-to-urban population shift, employment opportunities continue to be scarce in the countryside. Evidence from a wide variety of studies at the community and regional level strongly suggests that, for rural households, the decision to allocate some individual or individuals to the migratory labor pool is made largely as a result of an assessment of locally available income-generating activities. In brief, most research has concluded that, for many families, income-generating activities in the local community are so limited that extra resources must be sought outside the area. Clearly, then, a push factor stimulates the migratory process. Furthermore, the principal impetus to that push in Mexico is the lack of viable opportunities in farming. For a variety of reasons, agriculture has not been an effective "engine" for rural development in much of Mexico.

First, although most of those who migrate list agriculture as their principal occupation in the sending community, farming in sending regions often fails to provide adequate income for local families. Population pressure on the land in Mexico is severe, with an average of about 1.9 hectares of cultivated land available to each person considered to be economically active in agriculture in the 1970s (SALA 1983:47, 75). According to the agricultural census of 1970, 22.1 percent of all farmers in Mexico had farms of one hectare or less, and 42.4 percent had farms of between one and five hectares (Yates 1981:150). Most of these small farms are concentrated in the central plateau region. One study found that 61 percent of the production units in this central region could be classified as infrasubsistence, that is, incapable of providing for the subsistence needs of the farm household; an additional 15.3 percent of the units were classified as subsistence (CEPAL 1983:119). Fragmentation of landholding through inheritance patterns has also increased the pressure on land (Rivière d'Arc 1980:190).

Moreover, the land itself is often poor. Population increase, a harsh climate, and centuries of pressure on land have meant that much of the area suffers from extensive erosion. Yates (1981:43, 46–47) indicates that arable area in the country as a whole declined in the 1960s and that marginal land in the central region continues to be abandoned. An analysis of the 1970 agricultural census carried out by the Economic Commission for Latin America (ECLA) concluded that only 21.7 percent of all agricultural units could be made significantly more productive through the introduction of better technology. An additional 41.7 percent of the units were suitable for subsistence agriculture with the introduction of better technology over the short or long term. A final 36.6 percent of the units, however, could not be made productive enough to sustain even subsistence farming (CEPAL 1983:245–249). High outmigration regions also depend on rainfall for agriculture. Although rainfall is adequate in some regions, for much of the central plateau it is both inadequate and extremely erratic. In many areas, there is sufficient rainfall during four or five years of every ten; in other years, drought is a major threat; in 1982 drought destroyed about 40 percent of the crops in rain-fed areas (*LAM* 1986:74). The risks involved in farming such marginal lands are therefore great.

Nor do the crops grown on these lands provide much income or employment. Primarily dedicated to corn and beans, small farms in central Mexico are unable to generate more than subsistence or subsubsistence income (see CEPAL 1983:151–159; Cross and Sandos 1981: 66). The government has, since the 1930s, regulated the price of corn, and this price has been kept down to protect the low cost of staples for urban workers. Between 1965 and 1972, for example, the guarantee price of corn was frozen, whereas the consumer price index rose by 23 percent. In subsequent years, the guarantee prices for corn, beans, and other products have been raised, but when adjusted for inflation, they have demonstrated a long-term declining trend (see Table 6). Furthermore, marginal producers could not count on receiving the official price, as quality standards favored larger producers, transportation costs to government collection centers could be high, and government personnel were often less than honest in weighing and paying for the

Table 6. Guaranteed prices of basic staples, 1960–1982 (1970 pesos)

Year	Wheat	Beans	Maize
1960	1,290	2,119	1,130
1961	1,247	2,391	1,093
1962	1,211	2,321	1,061
1963	1,173	2,249	1,208
1964	1,111	2,129	1,144
1965	951	2,081	1,118
1966	915	2,002	1,076
1967	890	1,947	1,046
1968	869	1,900	1,021
1969	836	1,829	982
1970	800	1,750	940
1971	755	1,652	888
1972	711	1,556	836
1973	686	1,694	946
1974	834	3,851	963
1975	971	2,634	1,054
1976	812	2,319	1,085
1977	729	1,778	1,031
1978	792	1,904	884
1979	760	1,964	882
1980	699	2,362	876
1981	704	2,447	1,002
1982	552	1,581	706

Note: Deflated using implicit GNP deflator; base 1970.
Source: Mexico, SARH, *Determinación de los precios de garantía para los productos del campo,* 1984, reproduced in Sanderson 1986:203.

staples produced by peasants (see Grindle 1977a). Costs of production on marginal lands are high and increasingly require purchased inputs of seeds, fertilizer, insecticides, and at times hired labor (see Table 7). As a result land is often rented out, legally (for private land) or illegally (for communal land prior to the 1980s) to more prosperous farmers. Cross and Sandos (1981:31) report that, in a community in Michoacán, 80 percent of the ejido land was rented in 1967; in an ejido in Guanajuato, 30 percent of the land was rented. The poor potential of peasant agriculture is evident from a study of migrant farm laborers in Arizona which revealed that most were owners of ejido or private plots who either had abandoned cultivation as unremunerative or were unable to increase the technological inputs that would make the land more productive (Sánchez and Romo 1981). Labor migration from a region of Oaxaca was essential, Stuart and Kearney (1981) discovered, because "local peasant agriculture provides less than half of the staples consumed in the village and opportunities for cash income are woefully insufficient to fully supplement agricultural income" (p. 1).

Table 7. Costs and returns from maize production on medium-sized irrigated farms, seven municípios, guanajuato, Mexico (1974 pesos/hectare)

Item	1954	1974
Value of production	1,668	2,696
Purchased inputs	359	1,461
Net returns	$1,309	$1,235

Source: Roberts 1984:19.

Producers of staples make little money for their efforts and also find that the production of corn and beans under rainfall-dependent conditions provides little full-time employment. Yates (1981:235) estimates that the farm labor force in Mexico is as large as 10 million people during harvest periods and declines to about 3 million during the least active periods. In many communities, there are opportunities for agricultural work only three to four months of the year.[4] A major exercise in modeling agricultural production in Mexico estimated that, in 1968, the average hectare of rain-fed farmland provided only 1.11 person-months of employment per year, only 3.88 person-months per farm. In contrast, irrigated farmland provided 3.58 person-months of employment per hectare and 13.41 per farm (Bassoco and Norton 1983:136). In one region of a north-central state of Mexico, ejidatarios worked only fifty days a year on their plots; in a richer agricultural region of the state of Michoacán, they devoted 111 days to their plots, and in another region, 167 days were dedicated to work on the ejidos, producing average earnings of only $316 (Cross and Sandos 1981:31). On average, researchers found that underemployment was experienced by 86 percent of small landholders (Cross and Sandos 1981:68). In a comparative study of four agricultural regions, farm labor demand varied from 118 person-days in the poorest region to 75, 37, and 22 person-days in the remaining three areas (Roberts 1982:304). In another study, corn farmers in seven counties in Guanajuato dedicated 38 days per hectare to cultivation of this crop. These farmers, however, had already

4. These data should not be interpreted as an indicator of unemployment or underemployment, given the importance of off-farm economic activities in rural areas. Gregory criticizes data analysis of unemployment and underemployment in only one sector. "By treating the agricultural sector in isolation, one can uncover a huge surplus of labor. However, what may appear to be a surplus to the agricultural sector may not be surplus to the economy as a whole. To the extent that agricultural households divide their time between intra- and extrasector employment, the latter cannot be ignored" (1986:138–139). This pattern of employment divided among sectors is characteristic of Mexico's rural areas.

replaced much of their maize cultivation with sorghum, which required only 21 days per hectare (Roberts 1984:23).[5]

In some of the more favorable zones for agriculture, farmers have turned to more remunerative crops than corn and beans. Where they have done so, the nature of the crops grown has often provided greater income to rural families and more employment possibilities. Irrigated farmland in the central plateau area, for example, was estimated to provide 4.52 person-months of employment per hectare in 1968 (Bassoco and Norton 1983:137). With irrigation, double and even triple cropping is possible. In fact, in one region of the highlands of Morelos, tomato cultivation has resulted in a labor shortage that is met by seasonal inmigration from surrounding lowland areas. The results are not always straightforward, however.

Given the requirements of big labour inputs for the growing of such new cash crops as tomatoes, agricultural tasks—which previously could be done by members of the household—now demand the hiring of labourers. This demand for labour has meant that many local people who had migrated have come back to the Highlands, and it has also attracted people from other areas. The local population has grown, and this growth has three consequences: the competition for dry season labour increases, hence there is a vital need to use the local land for cash production; there is more pressure on local land and more competition for it. Cash crop production has also meant the introduction of technological innovations which favor increasing yields: improved seeds, chemical fertilisers and insecticides. This technology costs money. Access to cash is a requisite for cash crop farming. The heavy debts which many farmers incur often cannot be paid because of the risky nature of the specific crops they have chosen to grow, and because small-holding, cash crop farming is inevitably risky. Many people sell their land in order to pay their debts. This fact, together with differential participation of people in other activities, is leading to the emergence of differential social categories defined primarily in terms of access to cash and resources. [de la Peña 1981: 14–15]

Even in this case of considerable labor intensity, portions of the agricultural cycle continue to require little or no labor input. In other areas, the shift to fruit and vegetable production has brought with it a concentration of landownership through production contracts (on ejidos where land could not be bought or sold) or the mechanization of

5. In the region studied, the amount of land planted to corn declined by almost 50 percent between 1954 and 1974 at the same time that sorghum, uncultivated in 1954, came to account for 41 percent of the land in cultivation (Roberts 1984:15).

other crops such as cotton; both have resulted in the loss of agricultural jobs (Dinerman 1982:31). A survey of a large region in central Mexico revealed that, in 1974, the average number of days worked locally by the entire household (which averaged 5.4 members sixteen years of age or older) was less than 200 a year (Roberts 1984:35).

The employment opportunities provided by farming activities are limited, and in general, the financial rewards are even more limited. Many peasants find it better to abandon the cultivation of marginal land than to continue to try to eke out a subsistence living. Dinerman (1982: 30–31) reports that, in eleven states of north-central Mexico, more than 2 million hectares of land were abandoned between 1965 and 1974. In addition to a shrinking and marginal resource base, population pressure has meant that, over time, more of the rural population lacks access to land. In 1975, 48.9 percent of the agricultural work force was categorized as wage labor (Lassen 1980:161). Roberts (1984:37–38) discovered a 50 percent increase in landless wage laborers between 1950 and 1970 in one region of central Mexico at the same time that the total agricultural labor force increased only 10 percent. In an ejido in the state of Mexico studied by DeWalt (1979:51), half of the households had no land. In a community in Michoacán, Dinerman (1982:45) reports that 60 percent of all households had no land.

For the landless, opportunities for wage labor are limited in much of rural Mexico. Cross and Sandos (1981:31–32) indicate that, in the large sending region they studied, opportunities for landless laborers declined from 190 working days in 1950 to 100 days in 1960, and their income dropped by 18 percent in the same period. Further losses of jobs were reported for the 1960s. Wage labor opportunities in DeWalt's (1979:52–53) study were highly seasonal, and access to jobs was very competitive. In Oaxaca, little wage labor even on a part-time basis was available in one community; even forest resources were insufficient to provide more than very occasional income from cutting, hauling, and selling firewood (Stuart and Kearney 1981:6). Wage rates for agricultural labor are generally set at levels below the minimum wage specified for a region (see, for example, Stuart and Kearney 1981:6).

Nonfarm employment opportunities are also limited. In the Indian community of Ihuatzio in the Pátzcuaro region of Michoacán, Dinerman (1982:32) discovered that, by the 1970s, the rural use of industrially produced consumer goods had severely limited the possibility of absorbing household labor in traditional handicrafts. Clothing, furniture, sandals, basketry, and pottery that used to be produced in the village were now being purchased for cash from urban suppliers. Commercial pursuits, such as ownership of a truck or sufficient capital to stock a small store or to become a moneylender (or a combination of all three), provide some people with the means to generate often significant income in rural villages in much of rural Mexico. Production of local handicrafts for the tourist industry can also provide some employment possibilities. Nevertheless, as Stuart and Kearney (1981:7) argue in the case of Oaxaca, intense competition and the need for capital to pursue any of these pursuits limits the number of people who can engage in them. In fact, as we shall see, remittance income from labor migration is one source of the capital needed to begin such enterprises. Taylor's findings in two communities confirm the stark nature of local opportunities.

> Village employment rates are severely depressed by the small size of most household plots, by low per-hectare productivity, and by the seasonality of farm work. Fishing and wood-gathering provide some supplementary employment. Overfishing and rapid deforestation of the areas surrounding the two villages, however, seriously limits employment in both these activities. Finally, limited markets in the two villages and an underdeveloped marketing infrastructure to the outside place severe constraints on the profitability of local handicraft work. [1984a:66, 68]

Similarly, he discovered that employment did not increase with time spent in the community each year, another indicator of the poor local opportunity base (see Taylor 1984a:69–70, n. 2).

There are differences among communities and among migratory households in terms of whether migration is stimulated by desperation or whether it is seen as a strategy to advance the economic position of the household. Cornelius (1976b), in a study of nine communities in Jalisco, indicates that the decision to migrate "seems to be prompted in most cases by sheer economic necessity, rather than the desire to accumulate capital" (p. 25). Similarly, Sánchez and Romo (1981:4–5) refer to "dire economic conditions in Mexico" as a reason for labor migration to the United States, and Stuart and Kearney (1981:7) conclude that "migration is the only alternative to starvation" for many.

> For many households, a yearly trip to Culiacán is seen primarily as a means of household maintenance rather than a source of savings. After finishing the maize harvest in January, these households regularly depart for Culiacán and remain there until April or May, returning to begin planting. Though they may return with no savings, the family has managed to eat for four or five months without using maize stored from the harvest, and has

bought clothing. The stored maize remains to sustain them through the planting and weeding seasons, and some family members may return to Culiacán between the weeding and harvest and remit earnings to those who remain behind. [Stuart and Kearney, 1981:15]

The poor resource base in this community is apparent from the fact that 85 percent of the men and 55 percent of the women had experience as migrants (Stuart and Kearney 1981:7).

Dinerman (1982:72), on the other hand, discovered greater possibilities for capital accumulation in the local community but also warned that capital accumulation could provide an incentive for further migration (see also Malina 1980:9–12).

> Most persons, given their choice, prefer to invest in migration rather than in maize cultivation, especially if the latter requires replacing one's own labor with that of peones [hired labor]. Huecorianos no longer engage in wheat cultivation, supplanting it with plantings of alfalfa. More and more, they use irrigated land to cultivate vegetables for sale rather than crops for home consumption. Ironically, households use their profits from these enterprises to finance further migration.

There are also differences among households in terms of migration as a permanent or temporary feature of the household income portfolio. Cornelius (1976b:24), for example, identifies "target" migrants who engage in labor migration in order to meet a specific economic crisis or need—a crop loss, a debt due, a wedding or funeral, purchase of livestock. There are also the "professional" migrants who routinely leave the local community each year for several months, often going to the same location in search of work. In these cases, earnings from labor migration have become an institutionalized part of the household income portfolio.

Clearly, then, a set of objective conditions exists in many rural areas in Mexico to stimulate labor migration—population pressure on available land resources, a poor agricultural resource base, discriminatory government policies toward agriculture, a limited set of opportunities for off-farm employment, and the need for cash to make up or supplement subsistence income. Numerous studies indicate that rural inhabitants are highly sensitive to the limited economic potential of their local communities (see, for example, Arizpe 1982; Cornelius 1976a; Malina 1980; Mines 1981; Stuart and Kearney 1981). Migration is seen as an option—perhaps the only option available—enabling rural households to sustain or add to their income.[6] Overall, given the objective conditions and opportunities available in the local community, the decision to migrate is an eminently rational one, especially when it is coupled with perceived opportunities to earn significant income elsewhere. It is, then, a sensible allocation of labor from the perspective of the rural household faced by severe local economic constraints.[7] Moreover, given the experience of labor migration in many households and communities, venturing forth in search of additional income does not entail heavy risk of failure, especially by comparison with the probable results of remaining in the local community.

Given the lack of opportunity in much of rural Mexico, then, why do not households migrate permanently out of these areas? As we have seen, of course, rural households have in fact voted with their feet in massive numbers, going to Mexico City and increasingly to other major urban areas in the country to reside permanently. Some 2½ million have moved permanently to the United States (Reynolds 1979:123). For a very large number of other rural households, however, labor migration is a means to maintain rural roots, not to leave them. First, ejidatarios forfeit rights to the land if the allotted plot remains uncultivated for two years. If ejidatario households wish to retain access to the land, some members must remain in the rural areas to cultivate it (see Dinerman 1982:50–51). Small private farmers continue to use their land to contribute to a diversified income portfolio. Farming households regularly use nonpeak agricultural seasons to find employment elsewhere (see, for example, Arizpe 1982; Jacobs 1983:50). Indeed, peak labor migration to the United States coincides closely with the agricultural cycle in central Mexico when migrants leave in January, February, or March and regularly return for the harvests in the summer and again in the months of November and December (Cornelius 1976b:25; Stuart and

---

6. Taylor observes, for example, that "peasant allocation decisions reflect the dual objectives of ensuring the family's survival in the face of the uncertainties that characterize production and employment in the village economy on the one hand, and of increasing the surplus available to the household for consumption and investment on the other" (1984a:32). See also Conroy, Coria Salas, and Vila González (1979).

7. Roberts describes the impact of modernization in agriculture on many rural households in Mexico: "The household, when its size and composition allow, engages in a strategy of risk minimization through the allocation of its labour to different economic sectors and regions. The commercialization of agriculture increases monetary costs of production, the variability of farm income and, when combined with the substitution of manufactured for traditional consumption goods, the risk that this income will fall below a certain minimum level. The potential for local agricultural wage labour might also be decreased by changes in crop composition and mechanisation. Faced with these changes, the rural household will compensate by 'spreading risk across economic sectors and geographical space and securing alternative sources of income'" (1985:363).

Kearney 1981:15; Yates 1981:245). In this case, moving the entire household to the United States is often not feasible or desirable. Wages earned in the United States are very low by U.S. standards and inadequate to sustain a family in this country but are extremely high by rural Mexican standards and thus make significant contributions to household income as long as the household remains in Mexico (see Reichert and Massey 1980:489).[8] Moreover, although migrants regularly report their interest in the high wages to be found in the United States, they also express their disapproval of its lifestyle and declare it unsuitable as an environment for raising children (see Cornelius 1976a).

The risks involved in moving permanently to domestic urban locations are also considerable. Employment opportunities in Mexico's large cities have become notoriously poor, especially since the economic crisis of the early 1980s, when the petroleum bubble burst and the peso was devalued (see Wyman 1984; see also Dinerman 1982:31–32). Moreover, Cornelius (1976a:11–12) reports that labor migrants in his sample were well aware of the very limited employment opportunities in nearby towns and smaller cities. The risk of moving the household to a new location is thus greater than that of remaining in the rural communities. Kinship and locality ties and positive preferences for rural lifestyles are also important reasons why households remain rooted in rural areas, according to recent studies. Furthermore, as long as labor migrants retain a base in rural communities, they show a strong preference for returning to it and for helping maintain the household there. Several studies have argued that most migrants would remain in their local communities if economic opportunities existed there (see Barrientos et al. 1984:33; Rivière d'Arc 1980:191). Thus if there exist possibilities to reverse or limit the extent of labor migration from rural areas, then these migrants are the most likely to respond to new opportunities in the local area.

Who Migrates? Constraints on Portfolio Options

The conditions described in previous pages apply to much of the area of central Mexico where population pressure is great and local eco-

8. Gregory, for instance, reports that "a migrant employed at the U.S. legal minimum wage could earn in one day the equivalent of a six-day week's earnings in Mexico. For agricultural workers it would be the equivalent of 7.3 days of Mexican wages. Net earnings after deductions for social security and income taxes would reduce the size of the premium somewhat but would still leave it at a substantial level of five to six times the Mexican rate of earnings" (1986:197).

nomic opportunities are limited. Nevertheless, not all available labor migrates, nor indeed do all households or regions contribute equally to labor migration. Considerable research has provided us with a "profile" of labor migrant households. A number of studies make it evident that migration characteristics are linked to the composition of the household and its available options: some households are better able to take advantage of particular options than others, and some have a larger number of options available to them. Household size and age distribution appear to be critical factors here: larger households tend more frequently than smaller households to be migratory ones, and those with young unattached males are more frequently found among "sender" households; in this case, extended families are more suitable for engaging in labor migration than nuclear families (Dinerman 1982:51; Malina 1980:3–4; Roberts 1984:37). Ejidatario households send migrants more frequently than private smallholding households, artisans, or those with commercial pursuits (Cornelius 1976b:23; Dinerman 1982:51; Sánchez and Romo 1981:4–5). Households without land often have more than one member in the migratory pool at any one time.

Landlessness is frequently an important attribute of sender households, although it is not clear whether this category includes landless offspring of ejidatarios who are helping maintain the family plot with labor migration. If such is the case, landlessness may not be the most important impetus to migration, as suggested by de la Peña.

> It is nowadays not infrequent to find fathers who have only a small piece of private land (of, let us say, two hectares), or else a single *ejido* plot, to divide among two or three sons. In these cases, the whole family—even after the sons have married—sticks together, at least for a while—as far as agricultural tasks on the family plot are concerned. (They may, or may not, share the same household.) The father runs the farm, and divides the earnings among his sons and himself. Divided among several nuclear families, earnings are not even enough for subsistence. Wage labour has to be undertaken and, in some cases, becomes the main activity, while the family plot is just a supplement. In the end, one of the sons buys or rents the land from his brothers, or else the land is sold or rented to someone else. [1981:141]

The decision as to who among household members will migrate indicates important aspects of portfolio management. Demographic characteristics among migrants seem fairly consistent, as is evident in Table 8. A high percentage of them are young, male, without extensive

Table 8. Migrant laborer profiles

Study	Database	Age (years)	Education (years)	Percentage male	Percentage household head	Percentage landless	Percentage ejidatarios	Percentage private
Taylor (1984a)	2 communities in Michoacán							
U.S. migrants		27.5	4.05	60.0	1.0			
Domestic migrants		27.3	6.5	48.0	11.0			
Stuart and Kearney (1981)	1 community in Oaxaca							
U.S. migrants				100.0				
Domestic migrants				89.4				
Zazueta and Garcia y Griego (1982)	National survey							
U.S. migrants		54.7% less than 30 years of age	4.9	83.9	55.0			
Dinerman (1982)	2 communities in Michoacán							
Migrants						32.1	60.7	7.1
Nonmigrants						52.3	47.7	—
IMISAC (1982)	1 community in Michoacán							
U.S. migrants						66.0	32.0	1.0

education, and unmarried. When they are married, they often travel without their families. Women are more likely to migrate domestically, whereas males are the most frequent international migrants.[9] Domestic migration itself tends to involve more married persons and more heads of household.[10] Few migrants have regular employment in the sending community, but many have access to land. That is, as Cornelius states, "the outright unemployed, those who are simply unable to get any kind of even temporary employment at the local level, constitute a very small proportion of the total flow, with one exception: That is the young man just entering the labor force" (1981:23). This finding is confirmed in a national survey: although 20 percent of the migrants had not worked in the month prior to leaving in search of work, only 3.2 percent of all migrants were openly unemployed (Zazueta and García y Griego 1982: 77). Nevertheless, the extent of employment in the community is limited in terms of the number of available days worked. Taylor (1984a:66, 68) discovered that U.S. migrants worked only 50 percent of available days, and domestic migrants worked only 19.6 percent of the days they were in the local community.

Those who migrate tend to have characteristics that suggest they can contribute most to household income through their migration. They are young men and—to a lesser extent—women who are likely to be most eligible for the unskilled low-paying jobs that are generally available to labor migrants, both in the United States and within Mexico. Mines (1981:158–199) indicates that, for Zacatecas, males continue to migrate seasonally until their male offspring are of an age to need schooling in the agricultural skills that are important to the household's future access to land. Taylor (1984a:71–72) finds that migration significantly decreases with age. Moreover, where households have access to land, some available labor is usually allocated to agricultural pursuits, although wage differentials have become so great between the United States and Mexico that, where U.S. migration is a possibility, agricultural labor may increasingly be performed by women, who historically

9. Female domestic migrants also tend to spend less time in the community than male domestic or international migrants, an indication of the fact that they are frequently employed in more full-time positions as domestics, waitresses, or in other service jobs (see Taylor 1984a:70).
10. Taylor explains this difference for the two communities he studied. "Mexican destinations are well-suited for the short term, seasonal migration work. For a household head, particularly if he is a small farmer, the opportunity cost of domestic migration is small relative to that of migration to the United States, which generally requires a large commitment of both time and capital. Education appears to have a profound positive effect on domestic migration" (1984a:106). The implication is that education tends to encourage migration because there are no opportunities for employment of more educated workers in the village.

have not been cultivators in Mexico (Dinerman 1982:44). Elsewhere, hired labor may be employed for agriculture in order to free household labor for more remunerative migratory labor (Roberts 1982). In all cases, labor allocation appears to conform to a pattern of "maximizing use of available options and reducing risk" (Dinerman 1982:50–51; see also Jacobs 1983:51; Roberts 1982).

The data suggest that sender households are not generally among the poorest of rural dwellers, nor are they among the most well-to-do. For the first category, migration may be both too risky and too expensive for the household. Illegal migration to the United States, for example, where the best employment possibilities and the highest wages are to be found, often costs $240–$500, including transportation and fees for smuggling across the border (Cornelius 1976b:27; Reilly 1983:1). Some households, even with the aid of extended kinship networks, cannot afford this sum. Taylor's (1984a:60–64) study of domestic and international migration from two communities in Michoacán found that domestic migrants generally came from poorer households than U.S. migrants but that both categories claimed larger household incomes than laborers who remained in the village (see Malina 1980 for similar findings). The more well-to-do in the local community may not need to expand income through migration or may find sufficient employment at the local level so that they are not able to participate or are not interested in availing themselves of the opportunity to migrate.

Not all communities are as heavily involved in labor migration as others (Dinerman 1982:40; Kearney 1984:34; Rivière d'Arc 1980:189; Verduzco 1984:42). Once again, involvement appears to be clearly related to local opportunity structures. Poorer communities contain fewer households that can afford the costs of migration; they are often less integrated into regional or national market economies and thus feel less intensely the need for cash; potential migrants may also be more needed to contribute to mutual security networks in the local community (see Malina 1980:3–4).[11] In other cases, agricultural cycles may be longer and may therefore absorb more household labor. In addition, migratory networks are an extremely strong conditioner of who migrates and of where migrants go in search of work (see especially Taylor 1984a).

11. Some evidence from communities of indigenous groups (that is, those speaking an Indian language) indicates that an extremely poor resource base coupled with high levels of population pressure can stimulate extensive migration even from very poor and remote areas. Interestingly, migration to the United States is often as easy for inhabitants as domestic migration because of similar cultural distances from both receiving areas (Kearney 1984).

Migrant Destinations: The Pull Factors

Labor migration from rural Mexico is deeply institutionalized in local economic and social structures. Once again, most of the evidence for this statement is drawn from studies of U.S.-Mexican migration, but as is evident in Table 9, this is a large component of the overall migratory flow. Extensive labor migration to areas in the southern and western United States dates to the 1880s, when laborers were recruited to work on the construction of the railroads. Similarly, labor shortages during the 1920s and World War II led to specific U.S. programs to recruit Mexican labor for agricultural activities in the southwestern United States.[12] The Bracero Program (1942–1964) is the most recent of these officially sponsored programs. Indeed, in most of the local communities that have been studied, the flow of labor to the United States began or became well institutionalized under the Bracero Program in the 1940s and 1950s.[13] Much of the labor migration at that time was legal, and the preferred locations of California, Texas, and Illinois were established as beachheads for future migration. When the Bracero Program was terminated, the flow of migrants continued, although much of it became clandestine (Reichert and Massey 1980:487).

This tradition of labor migration in sending communities is extremely important in the destinations that are chosen by labor migrants. Migratory networks have developed that channel migration from particular households and particular communities to particular destinations, cushion the entry into the labor market of new migrants, and provide social support for the migrants and their families in the sending community. Mines (1981:13–14) refers to this phenomenon as a "migratory infrastructure" that, at its most developed, includes border settlements, "colonies" in the United States, and a series of job contacts (see also Cornelius 1976b:23–24; Kearney 1984; Stuart and Kearney 1981:1; Taylor 1984a). Similar linkages between places of origin and places of destination have also been found in domestic migration in Mexico (see Gregory 1986:108; Jacobs 1983:50). Kinship and friendship networks are most important, but employer networks are often also significant, so that particular employers encourage the cyclical return of particular

12. On the history of migration to the United States, see Cornelius (1981c); Mines (1981: 15–34); Erhlich, Belderback, and Erhlich (1979); Cross and Sandos (1981:1–15); Reisler (1976); Corwin and Cardoso (1978).

13. For an exception, see Kearney (1984:52). The Bracero Program provided for the legal entry of Mexican agricultural and other workers into the United States on the basis of contracts between the workers and the employers (see Cross and Sandos 1981:35–48).

Table 9. Migratory destinations

			Destinations	
Study	Database	Labor migrants	Mexico	United States
			percentage	
Shadow 1979	Villa Guerrero, Jalisco	822	31	69
Taylor 1984	2 villages, Michoacán	164	54.7	45.3
Dinerman 1982	2 villages, Michoacán	58[a]	43.1	56.9
Stuart and Kearney 1981	San Jerónimo Progreso, Oaxaca	101	65.3	34.7

[a]Households.

individuals or their relatives or friends. Active labor recruitment is also a factor in providing agricultural workers for harvest periods in zones of modern commercial production (see Astorga Lima 1985; Gregory 1986:108). Such well-institutionalized networks and recruitment techniques form a critical part of the "pull" factors in labor migration and appear to be more important than distance or, in an immediate sense, wage levels in determining where surplus labor is sent. These networks, evolving over time, encourage even greater durability of the migratory flow.

In addition to the kin, community, and employer networks, however, wage and employment structures play important roles in the determination of migrant destinations. This is particularly evident in movement to the United States, where wage differentials play a great part in attracting labor migrants. Cross and Sandos report that "no other two nations in the world sharing a common boundary have differences in average annual personal income that contrast as sharply as those of the United States and Mexico. Moreover, the disparities between California, the major receiving state, and Mexico are even greater than the aggregate differences between the two nations" (1981:59). In the 1980s, wage differentials in some parts of the border region were as high as 13:1. Indeed, important shifts in migratory destinations have occurred over time in response to wage differentials and job opportunities. Over time, for example, migration and the kinship and community networks that support it have shifted from agricultural jobs in states such as Texas and Arizona to service and manufacturing jobs in California. This state, with its relatively high wages, has now become the preferred destination of most migrants (Cornelius 1981b:11; Cross and Sandos 1981:38–39). Similar shifts in migratory destinations occurred after the peso devaluation of 1982. Factories in the northern border region of Mexico began to experience a labor shortage as workers began to prefer the limited risk of illegal entry into the United States and the promise of high wages to the alternative of secure employment with rapidly falling real wages in Mexico (Malina 1980:6–7; Miller 1984:6; see also Bustamante 1983:324 on the internationalization of labor).

Labor market demand in the United States has also changed over time, attracting migrants out of agricultural jobs into service and manufacturing jobs. In fact, changes in the labor market have made year-round jobs more available, and this trend may be important in encouraging more permanent migration to the United States, a trend that some observers have already detected (Cross and Sandos 1981; Maram 1980; Reichert and Massey 1980:488–489). There is also some evidence that migratory "sojourns" in the United States are becoming longer (Mines 1981:22).[14] It appears, however, that the impact of wage differentials and employment opportunities is mediated by the established kinship and community networks, which appear to adjust over time to such economic pull factors. This adjustment allows the household labor portfolio to be updated over time by incorporating more promising destinations while continuing to lessen the inherent risks of migration for the migrant. In consequence of such networks and labor market conditions in the United States, a national migration survey discovered that 77 percent of migrants found work within two weeks and only 18 percent were without work after three weeks in the community of destination (Zazueta and García y Griego 1982:86; see also Gregory 1986:110).

Migration and the Home Front

Extensive labor migration from rural homes and communities in Mexico has significant consequences for those households and communities. In fact, migration whose primary motive is to maintain the household in the place of origin is causing significant rural change. One important consequence of labor migration, for example, is a series of changes in farming practices. Rural households in Mexico that are able to invest in migration often send young able-bodied men in search of work. Where migration has become less seasonal and more oriented toward two- or three-year cycles, there have been changes in the avail-

---

14. Nevertheless, Conroy, Coria Salas, and Vila González (1979:2) argue that falling real wages in the southwestern United States between 1969 and 1978 and rising real wages in Mexico during the same period should serve as an incentive for temporary migration (given large absolute wage differentials) but as a disincentive for permanent migration.

ability of labor for agricultural tasks. Many households carry on as before, of course, because they are large enough to have sufficient labor for traditional crop cultivation, which, as we have seen, often does not require major inputs of labor except at harvest time. Some households substitute hired labor for migrating family labor. Other households have shifted to even less labor-intensive crops and are using more female and child labor for agricultural tasks. These experiences suggest again that, for many rural households, agriculture is not the primary basis of livelihood, even among those who own land. At the same time, there is little evidence that rural families are willing to give up their land even when it is not invested in as an important producer of income. Land continues to be highly valued by migratory households, perhaps as a bastion of minimal subsistence, even when labor allocation decisions distract from the income-generating potential of the land itself. Indeed, land markets appear to be increasingly competitive, with the price of land often bid up because of the inflationary pressure of remittance income (de la Peña 1981:129; Mines 1981:125–126). Similar pressures appear to be applied to construction material, reflecting the investment of remittance income in housing.

Perhaps the most disturbing aspect of the migration process for the local community is the fact that it seems to feed upon itself: labor migration stimulates dependence on migratory income and thus requires further migration. As Mines (1981:59) reports for one community, "The village has, in effect, become a rest, recreation and retirement center for current successful migrants, and a reproduction center for future migrants. Although many individuals gain by this process, the village economy remains frozen in its traditional low-productivity system." This Reichert (1981:64) calls the "migrant syndrome," in which migratory income serves as a critical source for the household budget but is not generally used for investment capital in the local area. Like Mines, he finds that the communities of origin "are places to return to in order to enjoy life, renew friendships, and strengthen family ties—not places where one earns a good living" (Reichert 1981:64). Even where remittance income has been invested in the household's development for the future—as in the purchase of livestock, a sewing machine, agricultural inputs, or a truck—further investment often implies more labor migration. That is, a common motive for additional migration among those who have apparently prospered through migration is to build up a stock of livestock, to purchase additional machines, to finance next year's crop, or to buy replacement parts (see, for example, Dinerman 1982:79–80). The strong implication is that many rural communities in Mexico are not strengthening the capacity for self-sustained development but are instead becoming dependent outposts of labor markets and economic opportunities elsewhere. They therefore pose a challenge to people concerned for the future of Mexico's rural development.

Conclusion: Can the Future Be Different from the Present?

There is little question that labor migration in Mexico is linked to limited economic opportunities in rural areas. Study after study has indicated that people leave their communities in search of income-generating activities because there exist few such opportunities in the local area or because it is the most economically attractive option available to them. Labor migration appears to be a reasonable choice for rural households in many parts of Mexico, offering them a viable means of maintaining themselves in their communities and in some cases making real improvements in their standards of living. Such migration also appears to present relatively low risks for the allocation of household labor—certainly involving much less risk than permanent relocation. The household remains in the rural community, adjusting its other income-generating activities and lifestyle to the absence of the migrant. As we have seen, certain households are more willing and able to engage in labor migration than others, certain communities have greater need for cash income and greater access to migratory networks than others, and certain regions are more likely to be marked by heavy outmigration than others.

A poor local opportunity base and a need for cash income are the most important factors stimulating labor migration from rural areas in Mexico. These economic conditions have a long history and have encouraged an institutionalized process of migration and dependence on migration, indicating that it has been "habit forming" for many households and communities (Dinerman 1982:79; Taylor 1984a:2). The destination of migration appears to be primarily determined in the short run by established networks of migration and over the longer term by differential wage rates. The implication is that, as long as the objective economic conditions exist, as long as networks facilitate and encourage migration, as long as wage rate differentials remain significant, and as long as there is a market for migratory labor, it will be difficult, if not impossible, to alter the incentives or means of labor migration. Massive temporary labor migration is a symptom of the fundamental reality of rural economic conditions in the country.

The foregoing review of knowledge about labor migration from rural Mexico has important implications for policy. Temporary labor migration of the type described here results from a process of rural development that has increased household and community dependence on economic conditions and opportunities outside the local area. Increasingly, then, rural communities are losing their viability as economic units and are unable to contribute effectively to their own future or to that of their country. At the most basic level, the solutions to household and community dependence on migration must clearly focus on the rural roots of the problems rather than on sanctions applied at the destination of migration. The primary push factor—the poor rural opportunity base—must therefore be addressed if there are to be any alternatives to the need to migrate in Mexico.

3

Searching for Opportunities

Labor migration from rural Mexico has been a routinized and institutionalized aspect of household survival and development for decades. In recent years, as the poverty and lack of opportunity in rural areas have worsened and as economic crises at the national and international levels have reverberated in local communities, the incidence of temporary labor migration has greatly increased, to the extent that in many villages half or more of the labor force may be absent at any given time. Considerable evidence indicates that decisions to migrate are purposive acts, responding to household-based income-generating strategies that weigh the availability and potential of economic opportunities in the local area against those that require migration. Often the choice is clear. Massive temporary labor migration is an indication of severely limited options in rural communities, of a situation that does not provide sufficient income-generating opportunities for large numbers of families.

In the previous chapter, the concept of income portfolios was used to indicate how household labor is allocated among activities that can help sustain a family in its rural environment. A critical challenge for rural development in the future is to make more opportunities available at the local level for portfolio management and expansion. Through such a strategy, rural areas that now depend on temporary labor migration for their very survival have the potential to become more self-sustaining entities, able to participate in and contribute to a broader process of national development. The implication is that the principal goal of rural development should be the creation of employment and income-generating opportunities in agricultural and nonagricultural pursuits. This chapter explores a series of potential rural development options from the perspective of income and employment generation.

To be effective, the rural development activities considered here— agricultural development, land reform, infrastructure investments, rural industrialization, linking migration to local development—must of course have the potential to improve the local resource and opportunity base and create sources of income and employment. That is, they must be oriented toward making household income portfolio development possible through a wider and more attractive set of local options. A strategy for rural development, however, must meet other criteria as well. It must promise to be feasible in its requirements from the political and economic systems and efficient in its use of resources. Most observers, for example, can point to a broad set of investments and improvements in rural Mexican communities that could generate a better quality of life for their inhabitants. Realistically, however, many such possibilities are not feasible within the context of Mexico's development, constrained as it is politically, economically, and administratively. Moreover, public and private resources are always limited and are particularly constrained by domestic and international economic crises in the 1980s. Criteria of feasibility and efficiency in addressing the question of income and employment must therefore be central in any assessment of the potential of various activities to broaden the range of economic opportunities available to local inhabitants. This chapter examines options first for their potential to generate employment and income and second in terms of their potential in the specific context of Mexico's development.

Rural Development as Agricultural Development

Rural development specialists affirm that agriculture serves as an engine for rural growth and welfare through increased productivity, higher rural incomes, and greater demand for nonfarm goods and services. Because of the centrality of agriculture to the rural development strategies discussed in Chapter 1, an assessment of its potential for income and employment creation in central Mexico is a useful point of departure in this survey. For many years the promotion of growth in agriculture has been the linchpin of rural development efforts in the country. Data presented in Chapters 1 and 5 indicate the extent to which administrations beginning in the early 1970s committed public resources to improving the productivity of rain-fed agriculture.

Numerous programs were implemented to achieve this goal in the 1970s and 1980s—from an integrated rural development program, PIDER, to an effort to direct appropriate technology toward rain-fed zones, the Rainfed Districts Program, to a massive infrastructure and social welfare investment program, COPLAMAR, to a revenue-sharing scheme, CUC, and finally to a massive subsidy program organized around a national food policy, the SAM.[1] Even earlier, a major program to bring the green revolution to rain-fed corn and bean production, Plan Puebla, was initiated in the 1960s with substantial public sector support. Later, rural development, with a strong emphasis on raising levels of production and productivity, particularly of staple crops such as corn and beans, was a major priority of the administrations of Luís Echeverría (1970–1976) and José López Portillo (1976–1982). The administration of Miguel de la Madrid (1982–1988), faced by severe economic pressures and committed to austerity measures with direct implications for public spending, gradually decreased the proportion of the budget allocated to rural development from about 15 percent in the early 1980s to 8.3 percent in 1985. Because of public austerity measures, this was a significant drop in the amount available to the sector.[2] Despite such cutbacks in investment, public concern for rural areas continued to focus on increasing technological innovation in rain-fed areas in order to increase agricultural productivity. As indicated previously, however, the overall impact of these investments was small.

Not all of the expanded budgets for rural development were actually spent in the rain-fed areas, of course. The SAM, for example, probably

1. I will describe PIDER, the rain-fed districts program, and the SAM in greater detail in Chapter 5. PIDER, initiated in 1973, is a large, integrated rural development program focused on the identification and development of microregions. The program has stressed production-oriented investments and has been strongly supported by the World Bank (see especially Cernea 1979; Grindle 1982). The rain-fed districts program was the response of the Ministry of Agriculture and Water Resources to concern about lagging production of domestic foodstuffs during the López Portillo administration (1976–1982). It was initiated in 1977 and focused on technological innovation on small farms in rain-fed areas (see Grindle 1981). COPLAMAR, a program for marginalized rural groups and regions, was created at the same time and focused on the more remote areas considered to have limited development potential. Its principal goal was to extend social welfare infrastructure to remote areas and to groups, particularly indigenous ones, that had benefited little from Mexico's development. The CUC was part of a larger government effort to decentralize the administration of development programs in the country. It provided state governors with some authority to allocate development funds within the state. The SAM, or Sistema Alimentario Mexicano, was a large-scale strategy for production, distribution, and consumption of basic foodstuffs in Mexico, oriented toward ensuring self-sufficiency in these products. It was directed toward the crops grown in rain-fed areas (see Grindle 1985a; Oficina de Asesores del C. Presidente 1980; Spalding 1984).

2. The amount allocated for rural development showed a slight increase to 8.4 percent of the budget in 1986.

benefited large and medium commercial farmers as much as it did peasants, and the highly developed irrigation districts continued to receive the lion's share of the agricultural budget (see Grindle 1981, 1985a; Spalding 1984). In addition, poor planning, inappropriate projects, administrative weakness, and corruption played a part in lessening the impact of government spending in rural areas. Undoubtedly too, many rural communities did benefit from improved access to basic services, better infrastructure, and more attention from credit and technical assistance agencies, even though such benefits are not always apparent at the aggregate level. Nevertheless, given such large outlays of public funds and the limited impact they appear to have had on production and productivity, it is worth assessing the extent to which strategies focused on increasing agricultural production and productivity can be expected to affect the conditions that limit income generation in rural areas and make labor migration an attractive option for many households. In addition, we should consider what specific interventions in the agricultural sector are likely to affect the paucity of employment and income-generating activities that encourages massive labor migration.

There are reasons to be cautious about basing rural development strategies on the potential for agricultural growth in much of the country. The resource base for agriculture in much of the central plateau region of Mexico is poor. Erratic rainfall, extensive erosion, and poorly endowed soils are characteristic of much of this area. Although rural communities in the central region are not generally the poorest in the country, they are among the most intensively farmed. The agricultural growth potential of much of this part of the country reaches what Carlos (1981:9–10) calls "the upper ecological limits" of their potential. As we have seen, the ECLA study (CEPAL 1983:245–249) indicated that more than a third of all production units in Mexico do not have the potential to provide for subsistence, and slightly more than two-fifths of the total units can be expected to achieve only subsistence levels.

Given this poor agricultural resource base, small farmers frequently opt to invest time and effort in other income-generating activities rather than to dedicate themselves to increasing output on their plots. As we saw in the previous chapter, small farmers not only decide on allocations of available labor and capital within the farming system but may also allocate investment between agriculture and other activities with the strategic goal of maximizing income at the same time that they minimize risk to overall income. This may be one of the subtle lessons from Plan Puebla, the well-financed and well-known green revolution project for small farmers. In the case of Plan Puebla, one researcher discovered that investment in new technology and the increased management time its application required was not an attractive alternative for many farmers, given low guaranteed prices for corn and the possibility of increasing their income from nonfarm activities (see Redclift n.d.:16). Population growth and landlessness add to the expectation that, for many areas, agriculture-led rural development strategies will not be able to respond effectively to the income and employment issues that shape the behavior of Mexico's rural households.

Not all parts of Mexico's central plateau are so limited in agricultural potential, of course. Some areas present considerable opportunities for growth and productivity if well-planned investments are made; perhaps 20 percent of the units can be made significantly more productive, according to the ECLA analysis to which I referred earlier (see CEPAL 1983). One of the municípios described in Chapter 4 is an example of an area in which investment in agriculture has brought considerable growth in output. More generally, the expansion of small-scale irrigation, where feasible, appears to be significantly related to improvements in agricultural output, making double and even triple cropping of some products possible (see Silvers and Crosson 1980). In some of the richer, more ecologically favored areas, cash cropping of highly remunerative crops—strawberries, tomatoes, onions, and other fruits and vegetables—has brought greater incomes to rural households able to produce these crops (Roberts 1984). Labor absorption is also considerable with these crops (Gregory 1986:118).

A critical component of these success stories is the availability of water; with irrigation or sufficient and dependable rainfall, farmers have responded to price incentives and demand in both domestic and international markets. Livestocking also appears to have increased the incomes of some rural households. It is important to note that the winter fruit and vegetable market in the United States—to which Mexico is already a major contributor—should expand in the future, in part as a result of declining investment in agriculture in the United States (Yates 1981:39–40; see also Thompson, Amon, and Martin 1985:3–4). Similarly, urban demand in Mexico for fruits, vegetables, and meat expanded notably under the impact of the petroleum boom; eventual recovery from the economic crisis of the early 1980s should also increase this demand (Yates 1981:23–29). Careful attention should thus be given to identifying areas of potential growth and designing interventions that can spur local agricultural development, particularly with regard to developing irrigation and water management schemes. Supportive ecological conditions, accessibility to markets, the availability of sufficient sources of water, and a growing demand structure should be characteristic of areas and crops that can be singled out for the development of agriculture-led rural development.

How realistic is it to expect the creation of options in agriculture significantly to reduce the incentives to search for temporary employment through migration? As was argued in previous pages, the potential of agriculture-led rural development is limited for much of the central plateau because of a poor agricultural resource base. In addition, long-standing government policies hinder more productive agriculture in rain-fed zones, primarily through effective price ceilings on basic crops and subsidies that encourage mechanization (see Grindle 1977a; Montañez and Aburto 1979). The production of corn and beans, the most common agricultural activity of the rural poor, continues to pay poorly, even though the production of these crops has been of intense concern to several recent administrations. The most important reasons for the poor payoff in this case is government pricing policy, set with urban consumers in mind (see Timmer, Falcon, and Pearson 1983). To the extent that such policies inhibit peasant farmers from making significant incomes from agriculture, there is clearly room to consider developing alternatives to them.

The administration of Miguel de la Madrid attempted to deregulate the price of basic foodstuffs and to dismantle the government's marketing agency, CONASUPO (*LAM* 1986:74). For a variety of political and economic reasons, however, such deregulation is extremely difficult to achieve. Mexico's experience with setting prices for basic staples dates to the 1930s, and CONASUPO has been a large and politically important institution, with clienteles in both rural and urban areas. Despite economists' considerable concern that the pricing policies set by governments and implemented through agencies such as CONASUPO fundamentally distort the "signals" sent to producers about levels and forms of production, the Mexican state has long remained committed to the ideas that peasant producers are protected from exploitative middlemen through the setting of guarantee prices and that low-income urban consumers are simultaneously ensured a "living wage" through low-cost foodstuffs. Indeed, partial deregulation of the prices of some basic staples in 1985 and 1986 added considerably to the cost of a basic market basket of food for urban workers; political elites are justly concerned about the potential of rising prices to create unrest in urban areas of the country (*LAM* 1986:39). This possibility is much feared by policymakers in Mexico, particularly at a time when austerity measures and high inflation have severely reduced real wage rates in the country.

Even if prices were to be set by the market, however, greater production of corn and beans would not necessarily have any direct impact on the range of economic opportunity available to rural inhabitants or on the structure of labor migration. First, the crops that the government would most like to see peasants produce absorb little labor. Given the government's concern for the production of corn and beans, agricultural policies directed toward rain-fed areas are not likely to stress incentives to move out of their production into more remunerative and labor-intensive crops. Even with the more labor-intensive crops, such as the fruits and vegetables mentioned earlier, much employment continues to be seasonal in nature. It is also the case that much of the most promising land for fruits and vegetables is already under production in these crops. Moreover, as we have seen, temporary migration is adjusted to agricultural cycles and the need for agricultural labor; many households continue their farming activities at the same time that they allocate labor to migration. In two of the regions considered in the next chapter, labor shortages during peak agricultural periods occur regularly. Moreover, agricultural development could actually increase the need for migration; in the community-level studies of labor migration, there is evidence that the use of technological inputs in farming, the use of irrigation, and the provision of credit may actually stimulate labor migration as households are pressed to generate cash for fertilizer and pesticides, for capital investments in agriculture, and to repay debts (see especially Roberts 1982). Finally, considerable evidence suggests that livestocking actually frees labor for migration.

I do not mean to argue that the Mexican government should not invest in rural development initiatives directed at increasing agricultural production and productivity. Increases in productivity—in an appropriate policy environment—have the capacity to raise rural incomes and to fuel the demand for nonfarm activities that generate both jobs and incomes. Furthermore, the Mexican government continues to have real interests in increasing the production of basic staples to feed the country's burgeoning population and to avoid what it considers to be politically threatening "food dependency" on the United States (see Grindle 1985a). For these reasons, investments in rain-fed agriculture must continue and should particularly address issues of irrigation, appropriate technology, and marketing systems for small farmers, which have been identified as important weaknesses in current practices. The government will also need to be concerned with the pricing system for basic staples, which currently makes it difficult for peasants to recover the costs of production, especially when they invest in technological innovations. Moreover, incentives to mechanize agricultural production, which have been important for the large-scale producers in the country, should be studied, particularly with regard to incentives to import and realistic prices for energy consumption. Actions related to pricing policy, of course, are likely to be those most beset by political problems, especially if significant reforms are implemented.

In assessing the future of agriculture-led rural development, it should be borne in mind that such a long-term strategy for agricultural development is costly, as recent Mexican experience demonstrates. Budget allocations for rural development fell in the 1980s, decreasing by 13 percent between 1983 and 1985 (*LAM* 1986:75). Moreover, large sectors of the rural population—those without access to land—will not benefit directly, if at all, from such investments. Furthermore, and most important for our purposes, there is the realization that such a strategy is unlikely to address directly the rural employment question and may even stimulate greater reliance on migration. Yates argues that the demand for on-farm labor in the future in Mexico is more likely to decline than to increase.

> There exist certain opportunities for increasing the volume of on-farm employment by means of specific attacks on productivity improvements in particular localities, and more generally through a policy of small-unit livestock farming. But simultaneously the steady march of mechanization will continue uninterrupted (even if the government were to desist from encouraging it with subsidies), and this trend will offset, probably more than offset, any employment gains through productivity improvement. [Yates 1981:240]

Similarly, Schumacher (1981:27–28), in assessing rural development expenditures under the López Portillo (1976–1982) administration, indicates that, although they had some impact on temporary job creation, they had a minimal impact on creating permanent jobs. Thus, he states, "the annual labor of ten workers in federally funded projects in rural areas is necessary to provide one permanent new rural job." His results and estimates, reproduced in part in Table 10, indicate both the high cost and the limited impact of employment generated through investments in agriculture-led rural development. Nevertheless, as suggested by Yates, to the extent that the government invests in agricultural

Table 10. Job creation impact of public investment in agriculture, 1971–1982

Type of job	Annual average		
	1971–1973	1975–1977	1980–1982
Permanent			
Cost per job, pesos	312,500	621,000	690,000
Total man-years of employment	15,400	29,500	111,884
Temporary			
Cost per job, pesos	53,000	95,000	95,000
Total man-years of employment	90,566	192,000	812,000

Source: Schumacher 1981:27–28, based on estimates derived from data from the Ministry of Programming and Budget and the Ministry of Agriculture and Water Resources.

production supports that might have some impact on labor migration, targeting of investments for specific communities and particular projects may have some payoffs.

Altering the Structure of the Ejido

One of the most significant structures to emerge from the violent Revolution of 1910 in Mexico is the ejido, a communal landholding granted by the state to a group of legally incorporated farmers, or ejidatarios. An agrarian reform, first promulgated in 1915 and then incorporated as Article 27 of the Constitution of 1917, was massively implemented under the presidency of Lázaro Cárdenas (1934–1940). Succeeding administrations continued to distribute land, often on a piecemeal basis and generally with political rather than developmental objectives. To receive land, groups petition the state for title, and once this has been granted (often after years of procedures and litigation), a committee of members administers the affairs of the ejido. Individual ejidatarios are assigned plots of land for themselves and their heirs, subject to fulfillment of certain obligations to the ejido. Ejido plots can be cultivated either collectively or individually by the ejidatarios; the vast majority are currently worked on an individual basis. Ejidos currently account for more than 40 percent of the agricultural land in the country and number some twenty-two thousand. There are approximately 2 million ejidatarios, and with their families, they account for the largest single category of farmers in the country.

Many people concerned about the underproductivity of Mexican agriculture point to the poor performance of the ejido sector and the lack of incentives for increased production that this form of land tenure

Table 11. Comparative characteristics of ejidal and private farms in Mexico, 1970

Type of farm	Irrigation as percentage of arable land	Value of crop output per arable acre (pesos)	Value of livestock output per animal unit (pesos)
Private	24.4	—	721
Over 12.5 acres	—	589	—
Under 12.5 acres	—	514	—
Ejidal	15.1	393	284

Source: Yates 1981:71, 134, 135.

purportedly encourages. There is empirical support for this perspective. On a wide range of measures, ejidos consistently underperform in comparison with private farms. Between 1950 and 1970, for example, private farmers produced, on average, twice as much per unit of land as did ejidatarios. During this same period, the gap between productivity on private farms and ejidos widened as private farmers increased their output more rapidly than did the ejidatarios. Similarly, output per person on private farms was estimated to be four times greater than that on the ejidos (Yates 1981:161). Even small private farmers, those with under five hectares of land, who are frequently categorized as subsistence farmers, generally outproduced ejidatarios in the value of their output per unit of land (see Table 11). Thus, it is argued, the ejido structure stands in the way of more efficient production and consequently of higher levels of rural wealth and welfare. In particular, the group basis of the landholding and the fact that ejido land cannot be sold or mortgaged is seen to be an impediment to the development of entrepreneurial talent in farming and the effective performance of land and credit markets. Accordingly, critics have argued that greater incentives would come from opening up the ejido to land, labor, and credit markets so that ejidatarios could reap the full benefits of their investments and risks. By this perspective, changing the structure of the ejido would do much to increase the productive potential of the land in Mexico and the incomes of farm families. In fact, in the early 1980s, the government experimented with such alterations, attempting to make ejido land and labor more accessible to rental markets.

Alteration of the structure of the ejido is therefore another rural development strategy that should be considered for its potential to increase the extent of economic and employment opportunities in rural communities in Mexico. For our purposes, proposals to change this important category of land tenure in Mexico must demonstrate some capacity to promise improved income-generating opportunities in rural areas. Proposals must also be feasible within the constraints of Mexico's development. An application of these criteria to proposals to alter the structure of the ejido suggests that it is not a viable option for Mexico in the 1980s and 1990s. In fact, a review of the discussion of household income portfolio management in Chapter 2 indicates that altering the structure of the ejido might actually increase the extent to which rural inhabitants become dependent on income generated elsewhere and the extent to which they abandon agriculture altogether. Our evidence indicates that the ejido structure is important in keeping households rooted in rural areas. Ejidatarios are frequently involved in temporary labor migration, but they structure their migration around the need to maintain access to the ejido plot and to ensure that they are able to leave it to their heirs. Legally, they are required to keep it under cultivation at least one out of every two years. Moreover, the greatest number of labor migrants come from landless households; in the event that ejido land is made more subject to market forces, land concentration and consequent landlessness is a likely result. Thus, altering the structure of the ejido could easily increase the incidence of labor migration, and there is little evidence that such a change would create the kind of economic opportunities that would provide more local options to those who engage in temporary labor migration.

In addition, it should be recognized that the failures of the ejido sector are not necessarily those of lack of entrepreneurial talent among ejidatarios. Much of the sector's disappointing performance results from the poor natural resource base in the regions where the large majority of ejidos are established and of discriminatory government policies. Ejidos with access to irrigation, technical assistance, and credit have generally performed comparably to private farms with similar endowments (see Hewitt de Alcántara 1976:187). Moreover, as we saw in the previous chapter, government investment in agriculture has been heavily skewed toward the zones of large-scale and irrigated agriculture where the number of ejidos is small. Since 1940 government policy has sought to promote "capitalist" farming as the sector with the greatest potential rather than the "socialist" ejido sector (see Hewitt de Alcántara 1976: 192–200; Sanderson 1981). Credit, technical assistance, improved and appropriate technology, market facilities, and other services have been directed away from the heirs of the agrarian reform toward zones and groups thought to have greater productive potential. Therefore, in many ways, measuring the ejido sector against private farms obscures the extent to which development of the one sector has been subsidized by lack of attention to the other. Much more than tenure structure is responsible for the poor showing of the ejidos. Even if we grant that the ejido sector has been systematically discriminated against by government policy, however, it can be argued that reversing this process is too costly to be feasible. Moreover, given the poor natural resource base of a large number of ejidos, we may ask whether many could actually become more productive even if they had greater access to technical innovation, credit, and a wide variety of agricultural services supplied by the state.

The most telling limitation on the potential of restructuring to improve the rural opportunities available to ejidatario families, however, is straightforwardly political. There are major political constraints on the extent to which the structure of the ejido could be altered, constraints that reach back to the Revolution of 1910, the Constitution of 1917, and the massive agrarian reform of the 1930s. The government of Mexico has made a political commitment to its peasant population through the ejido structure. The enduring political stability of the country is due, in no small part, to the creation and maintenance of that structure, however underproductive it may appear to be. The ejido structure will therefore continue to be a central feature of Mexico's rural areas, particularly in the rain-fed zones where the vast majority of ejidos are now located. Efforts to alter this structure in any significant way are thus not likely to have political success. It is therefore not to be expected that such efforts will hold much promise for the creation of more economic opportunities in rural Mexico.

An alternative solution would be to expand the ejido structure. Development specialists concur that rural development strategies based on equitable access to land and other resources are likely to lead most effectively to dynamic economic growth and improved welfare. Moreover, in Mexico, if access to an ejido plot tends to anchor rural families more firmly in agricultural communities, then expanding the sector should result in at least a reduction in permanent outmigration to overcrowded and overextended urban zones in the country. Expansion might even have a small impact on temporary labor migration if it reduced the number of landless laborers, the most frequent migrant category. In an exploration of this "what if" scenario; Silvers and Crosson (1980:19) concluded that the expansion of the ejido system would likely have a slight but not important impact on restraining rural-to-urban migration. Certainly an expansion of the agrarian reform could bring greater equity to rural Mexico, where the biases of the country's development process are most apparent.

Nevertheless, the feasibility of expanding the ejido structure in order to increase economic opportunities in rural areas is problematic from an economic perspective. As we have seen, agriculture-led rural development on lands of limited productive potential—a strategy implicit in the ejido restructuring alternative—is costly in terms of the inputs of money, time, and administrative capacity it requires from public coffers. Expanding the agrarian reform cannot be expected to have a positive impact on household income or economic opportunities unless it is accompanied by massive investments in infrastructure, credit, technological innovation, social welfare services, and other resources made available to current and potential ejidatarios. Lack of access to such services and controlled prices for basic products have been important causes of the poor performance of Mexico's ejido sector in the past and will continue to discourage increases in productivity on such farms. Improving the rural economy through this option should therefore be considered by governments that have not only the administrative capacity and political power to implement an expanded agrarian reform but also the financial resources to commit large amounts of money and professional talent to improving production on rain-fed and often marginal plots. Unfortunately, it cannot reasonably be expected that for the 1980s and 1990s the Mexican government will have available the resources to accompany a renewed agrarian reform with the level of investments needed.

There are other reasons to suspect that a major initiative to expand the size of the ejido sector would not have a significant impact on the rural conditions that encourage high rates of temporary labor migration. To begin with, much of the land in the regions from which most of the temporary migrants come is already incorporated into ejidos. Second, there is precious little newly available land to incorporate into ejidos in areas of high population pressure on the land (see Yates 1981:42–67). Again, however, the most important factor arguing against the feasibility of expanding of the ejido sector is political. Although individual peasant communities continue to press for the creation of new ejidos or the legalization of outstanding title claims, there is little political support at the national level for "ejidoization" and the facts of economics and population pressure provide little incentive for pursuing such a path. On the contrary, expansion of the agrarian reform is not viewed as an option by most public policy makers in Mexico and has been explicitly recognized as such by recent presidents José López Portillo and Miguel de la Madrid (Grindle 1985b; Sanderson 1984). The agrarian reform of the 1930s was possible because of committed political leadership, an organized and mobilized peasantry, and a weakened sector of large landowners. By the 1970s and 1980s, political leaders had become uninterested in agrarian reform, the peasantry had become largely coopted and controlled by the dominant political party, and large landowners had become powerful politically and economically through the successful development of commercial agriculture.

Infrastructure Investments

Roads, market and storage facilities, irrigation works, electricity, communication linkages, and social infrastructure such as schools, sanitation, and health facilities are important ingredients in rural development and can have important consequences for the quality of life that is available in rural areas. They can also stimulate economic ventures that generate jobs for rural inhabitants, such as investment in small enterprises (see Mellor 1976; World Bank 1978). In addition, investments in infrastructure can encourage people to remain rooted in rural areas if the often wide disparities between rural and urban standards of living are modified. They can make agricultural production more remunerative, particularly when irrigation facilities are expanded (see, for examples in Mexico, the findings of Bassoco and Norton 1983:137, Silvers and Crosson 1980:119). Infrastructure development is therefore often recommended as a central component of rural development strategies.

Infrastructure can be created through rural public works projects, which are thought to have considerable potential to create employment opportunities (see Thomas 1974). Not surprisingly, governments concerned about the potential for rural upheaval during periods of natural disaster or peasant mobilization have often turned to rural public works projects as a means of maintaining the social peace. Most experience to date suggests that rural works can have considerable impact on short-term employment opportunities and further that they can be planned and pursued to complement rather than to conflict with peak seasons for agricultural labor demand (see Thomas 1974). Schumacher (1981: 4–5) indicates that, in Mexico, rural works can provide a short-term solution to employment problems while agricultural and rural industry development schemes are put in place for more long-term solutions. In spite of their obvious benefits, however, it should be borne in mind that rural works programs are expensive and are often managerially intense, limiting their broad application to the problem of income generation in rural Mexico.

More generally, however, investments in infrastructure, unless carefully planned, can also stimulate increased labor migration. When remote and poor areas are made more accessible, for example, it may become easier for household members to join the migratory labor stream. Similarly, roads may bring with them greater penetration of the national and international market economies and may generate the need for households to have access to more cash income, often attainable only through labor migration. It should also be remembered that the heaviest migration comes from areas and households that are among the more advantaged, not the poorest ones. Thus if rural incomes rise and previously isolated areas become more developed, labor migration may become more affordable and possible. Given the fact that the most enterprising of rural inhabitants are often the first to engage in temporary labor migration and are also among those most likely to migrate permanently, there may be a "brain drain" effect on local communities. Such a consequence is most likely to result from investments in social and physical infrastructure that make it easier to take advantage of economic opportunities outside the local community and that do little to improve opportunities within it. Rhoda (1983:34), for example, presents data to suggest that "migration appears to be stimulated by interventions which increase access to cities, commercialize agriculture, strengthen rural-urban integration, raise education and skill levels, or increase rural inequalities." Silvers and Crosson (1980:146) found that rural-to-urban migration increased with access to education, and Cornelius (1976b:15) studied one community with so little economic potential that building a road "only hastened the community's extinction."

These are sobering findings for those concerned with creating a wider range of options at the local level so that labor migration is not the only alternative for many rural households. The reservations noted above, however, do not mean that infrastructure is an unimportant aspect of a viable rural development strategy. Infrastructure clearly is and will continue to be a critical factor in allowing rural areas to develop their potential in agricultural and nonagricultural economic activities. Some specific examples of the role of infrastructure in encouraging local development are presented in Chapter 4. These examples confirm the positive impact that such investment can have. Nevertheless, investment in infrastructure is not automatically beneficial to local communities; it has its most positive impact when it is planned around the creation of specific economic opportunities that build on local economic potential. In general, it can be proposed that infrastructure projects will have the most positive impact on rural development initiatives when they are geared toward or are coupled with the creation of specific income-generating activities in the rural areas. In isolation from such carefully planned and complementary investments, they may prove counterproductive to the longer-term development of the community. According to Cornelius, for example, "our research indicates that the most demographically destabilizing mix of public policies affecting rural communities might consist of social services, especially health care, and physical infrastructure, especially roads; all in the absence of efforts to stimulate new local jobs, especially the non-agricultural variety" (1976b:19–20).

Rural Industrialization

It is often recommended that rural industries generate significant employment opportunities in rural areas. Currently, in fact, rural development specialists consider the creation of nonfarm employment opportunities an essential part of making rural areas more developed and self-sustaining (see Mellor 1976, 1985; World Bank 1978). They link the need for such sources of employment to realities of current and projected population pressures. Johnston and Clark (1982:78), for example, argue that "expansion of nonfarm job opportunities is a necessary condition for the relative and eventually absolute reduction in the size of the agricultural population and labor force. In brief, changes in the occupational composition of the labor force and in the composition of output are crucial: they prevent or eliminate overcrowding on the land and make possible higher levels of productivity and per capita income." As we have seen, rural industrialization is described as an important ingredient in the successful development of rural areas in the United States, Taiwan, Japan, and Korea. Interestingly, moreover, many rural inhabitants in Mexico believe that this solution has potential for their communities. Cornelius (1976b:20) discovered in nine high migratory communities, for example, that "soil and climate, fluctuations in prices, and uncertain access to credit and fertilizer make agriculture too precarious in most of the communities. Moreover, the anticipated natural population increase cannot be absorbed by the agricultural sector, even with expanded land resources. When asked, What would have to be done to keep so many people from migrating? residents are quick to respond, Bring some industries here." A variety of similar responses from rural inhabitants of four municípios are reported in Chapter 4.

In rural development theory, rural industrialization is expected to result from the increased productivity of agriculture, which leads to higher rural income and demands for goods and services that can be produced or provided locally or increased ability to produce goods for urban and international markets (Johnston and Kilby 1975; Mellor 1976, 1986). In some of the regions of Mexico that are reasonably well endowed for agriculture, such as one of the municípios discussed in Chapter 4, the creation of linkages between agricultural development and nonfarm employment opportunities has considerable potential. As we have seen, however, many of the areas in Mexico that are most characterized by dependence on labor migration are areas with very limited, if any, potential for expanded agricultural production. It is therefore a question of the potential to create sources of nonfarm employment that do not depend on increased income from agriculture. This problem presents a challenge to rural development theory but also provides an opportunity for thinking creatively about alternative routes to the development of many of Mexico's depressed rural areas. Several possible strategies for rural industrialization are considered below in an effort to find alternative engines for rural development where the prospects for agriculture are poor.

Agroprocessing

Development specialists who have studied rural enterprises generally recommend a number of industrial activities that might be promoted to create employment and raise incomes. Most frequently, they suggest that agroprocessing offers considerable development potential. As a strategy to create nonfarm employment, agroprocessing can be linked to programs to increase the productivity of agriculture; the locally based linkage between primary and secondary sectors is regarded as a natural and self-sustaining process, providing considerable independent development potential for the local community. Chuta and Liedholm (1979:27) argue that rural processing activities have demonstrated considerable efficiency in Third World countries. Similarly, Bruton (1974:70) indicates that such processing can play an important role in stimulating employment in small towns in Third World countries. This form of rural industrialization is therefore recommended as the first and most natural step to take in creating more diversified and self-sustaining rural economies (see, for example, Austin 1981).

Indeed, agroprocessing industries have a considerable and successful history in Mexico, a country that began undergoing a long period of import-substituting industrialization in the 1930s and 1940s. By the 1970s, domestic markets had become well supplied with processed foods of Mexican origin. More recently, agroprocessing industries have become heavily involved in the export market and in many instances have been important in encouraging shifts in crops produced and in levels of agricultural production (Burbach and Flynn 1980; Feder 1978; Sanderson 1986). A wide range of fruits and vegetables are now grown and processed in Mexico, many of them destined for the U.S. market— tomatoes and strawberries are the best known examples. Livestock and dairy production have also been substantially industrialized in Mexico. Large multinational and national corporations have invested heavily in processing plants and have become heavily involved in production schemes through contract agriculture. Under this form of production, small producers and ejidatarios (who cannot lawfully sell their land) sign contracts with the large firms (who, if foreign, cannot by law own land), agreeing to deliver a stated quantity and quality of agricultural product to the firm, which agrees to purchase a given quantity and quality for a stated price. The firm often provides seed, technology, technical assistance, and credit, whereas the farmer provides land and labor for this arrangement (see Burbach and Flynn 1980:182–191; Goldsmith 1985; Sanderson 1986:52–54). Large areas of the more fertile regions of central Mexico are now involved in this form of contract agriculture. Other large firms are heavily involved in processing agricultural products for animal feed. Meat packing and dairy production have also increased steadily under the auspices of foreign investment from the United States, which in 1978 had come to account for $229 million, or 22 percent of U.S. foreign investment in Mexico for that year (Scott 1980:10). Both the domestic and foreign market and the success of agroindustries thus suggest that potential exists for creating jobs in this form of rural industrialization.

Nevertheless, there is also reason to be cautious about the increased employment and income benefits that can be expected from the promotion of agroprocessing in Mexico. One obvious limitation on its potential is that many of the most densely populated rural areas of the country lack the agricultural potential to produce crops for the agroprocessing activities. In these areas, low levels of production, erratic harvests, and poor crop quality severely limit the expansion of this form of rural industrialization. Even in areas where such investment would be more ecologically feasible, however, caution about the employment potential of agroprocessing expansion is in order because of the very sophistication that the industry has already achieved in Mexico. It must be remembered that this is not a country with a large unmet domestic demand—four decades of import substitution policies have stimulated a large, complex, and sophisticated industry and a relatively discriminating consumer market. The export market, dominated by large multinationals and often controlled economically by commodity brokerage houses in the United States, requires considerable capital, high-quality control standards, and sophisticated marketing skills and contacts. To promote further industrialization along these lines requires equally sophisticated planning and management and considerable investment capital. Yates, for example, concludes that the potential for agroprocessing may be limited to fairly large-scale and sophisticated firms. "Nowadays food processing has become a highly technical operation which could not be managed by inexperienced ejidatarios. It requires a steady supply of products for processing, not a few tomatoes one week and a few cucumbers the next, and finally to be profitable it has to be fully mechanized and offers only a small amount of employment. . . . there is little likelihood that rustic endeavors could compete against the well-established large enterprises" (1981:240).

Yates is perhaps overly pessimistic about the possibility of stimulating agroprocessing in rural areas; there may be products such as dairy and leather goods that can be stimulated with less difficulty (see, for example, Mellor 1976:173–175). Even here, however, the impact of such industrialization on rural labor markets must be scrutinized. The type and seasonality of employment created by agroprocessing, for example, is problematic. Such industries generally employ labor at periods of peak agricultural activity, when labor demand is also great. We have also seen that temporary labor migration is a phenomenon stimulated by lack of local employment during nonpeak agricultural periods, when rural households seek to diversify their income-generating portfolios. In general, agroprocessing is not likely to make major contributions to creating timely alternatives to temporary labor migration, although industries such as dairying may avoid this limitation. Similarly, small livestock production can overcome many of the problems of seasonality and marketing that characterize crop production and processing linkages. In particular, the poultry industry specializing in eggs or meat production has potential that can be developed in relatively small operations with considerable employment potential.

Opportunities provided by investment in agroprocessing can contribute to improving household income and might address the problem of outright unemployment, particularly among landless workers. To the extent that such activities contribute to raising rural incomes, they can make rural areas better markets for industrial goods and can generate greater consumer demand for goods and services that can be produced locally, providing an opportunity for local investment and entrepreneurship. The stimulation of agroprocessing should be assessed in terms of regionally based opportunities for its development and expansion and in terms of its contribution to employment.

Small-Scale Rural Industries

In addition to agroprocessing, development specialists also recommend small-scale rural industries for their potential to create additional sources of employment and income in local communities. In their extensive review of nonfarm employment experience, for instance, Chuta and Liedholm conclude that "rural non-farm processes are generally both more labor intensive and more productive per unit of capital than their larger, often urban based counterparts in the same industry" (1979:41; see also Anderson and Leiserson 1980; Berry 1984). In more recent work, they have examined the location and efficiency of small-scale production activities in a number of Third World countries and have discovered that the vast majority of such enterprises are located in rural areas and that a potentially wide range of small-scale industries can be found and further promoted (see Chuta and Liedholm 1984). Four main groups of activities—food processing, textiles and clothing, wood processing and working, and metalworking—appear to account for the largest amount of rural off-farm employment (Anderson and Leiserson 1980:233; World Bank 1978:25). Chuta and Liedholm (1984:300) present data to suggest that activities such as tailoring, dressmaking, rice milling, furniture manufacture, and baking enterprises have done well in many rural areas, as have services such as bicycle repair, automobile maintenance, and electrical repair work. Their review of current evidence indicates that less positive growth appears to have occurred in shoe production, leather working, and pottery; spinning, cloth production, and blacksmithing present mixed experiences. There has also been some positive experience with cement block and perfume production and with certain kinds of textiles in several countries.

The general argument for small-scale rural industries suggests that such activities can be directed toward rural markets, urban markets, or international markets for consumer goods and the large-scale industrial market for intermediate goods. Subcontracting may be a viable means to stimulate the linkage between small-scale production and larger-scale industry located in small, medium, or large urban areas (see Mead

1984). Where rural incomes are very low, of course, rural markets will also be constricted, but Chuta and Liedholm suggest that rural markets can present excellent opportunities when incomes are raised and the elasticity of demand increases (1979:23, 62). Most such activities tend eventually to become concentrated in rural towns (Anderson and Leiserson 1980:241). For a number of activities, small and medium-sized towns may be the most appropriate habitat, encouraging more integrated and balanced linkages between rural hinterlands and large urban areas (see Rondinelli and Ruddle 1978:58–60).

In the specific context of rural Mexico, small-scale industries also hold considerable promise to increase the options available to people who might otherwise have no alternatives to temporary labor migration. Rhoda (1983:51) argues that much of the development of rural enterprises would actually be based in small towns and market centers, where infrastructure and marketing potential are often available. This is a viable policy option, however, as such development, if located carefully, can provide jobs that are accessible to rural "commuters." In the municípios discussed in Chapter 4, a considerable amount of such commuting is in evidence. In addition, Rhoda (1983:57) indicates that rural enterprise development would have a significant impact on rural outmigration to large urban centers. Farming-out systems, with small or market towns as the base, also promise employment opportunities in specific industries such as the manufacture of clothing. An important aspect of their promise is that small-scale enterprises, especially when they are linked to the household, can often be adjusted to other labor demands within the household.

With small-scale rural enterprises, the problem of who would benefit from increased employment must also be addressed. Most small-scale industry tends to be family based and to employ primarily household labor. Such small enterprises offer the flexibility and minimal risk that is highly complementary to household income portfolio management. Still, there may be significant limitations on the income that can be generated unless small enterprises can expand beyond the household. Significant problems—lack of capital, procurement and marketing channels, or entrepreneurial and management experience—are often encountered when these family-based enterprises seek to expand (see Berry 1984; Chuta and Liedholm 1979:66–68). There is, therefore, considerable need to investigate the specific nature of the problems encountered by such enterprises in order to understand how they might be stimulated. In many cases, family-based production or service enterprises may be the most appropriate for rural needs and opportunities. In other cases, an economic environment suitable for stimulating the expansion of small-scale units and the managerial, technical, financial, and marketing assistance needed by local entrepreneurs are the most important "missing links" (see World Bank 1978). Unfortunately, little is currently known of the specific types of public policies that are most supportive of rural small-scale industry (see Berry 1984). At the project level, various interventions have been suggested, such as:

(a) vocational training programs to upgrade skills, improve the standards and conditions of training, and encourage innovation; (b) various banking and credit schemes to mobilize rural savings and improve access to working and investment capital; (c) trading services to widen markets and improve access to supplies of material and equipment; (d) research and technical assistance to further stimulate innovation and the adoption of appropriate technologies, and to ensure their effective delivery and use; and (e) rural industrial estates [World Bank 1978:8–9].

Industrial Decentralization

A third possible route toward rural industrialization is through the decentralization of large-scale industry. Most often, such decentralization is geared not toward specifically rural areas but to intermediate and small cities that are close to areas of high rural underemployment but that also offer the physical, economic, and social infrastructure important for larger industries. Mexico has a number of characteristics to suggest that industrial decentralization is a viable option. Over the period since 1940, the country has developed a large, often dynamic, and sophisticated industrial sector. Currently, this sector is heavily concentrated in large urban areas, particularly in and around Mexico City, where 50 percent of the country's industrial capacity is located. A large number of intermediate and small cities, however, can offer enough appropriate infrastructure to make decentralization attractive to government planners and private investors. On the basis of such reasoning, the logical policy recommendation is that the government should stimulate industrial decentralization to provide jobs and to create additional economic opportunities for rural inhabitants close to their homes.

Indeed, the theme of industrial decentralization has been of importance in national development plans at least since the early 1970s (see Alonso 1984; Yates 1981:25). These policies have had some impact: as of 1975, the highly industrialized Mexico City region had become somewhat more decentralized within that region; states surrounding Mexico City experienced increased industrial growth; and areas outside this region grew more rapidly than the central industrial zone (see Alonso 1984). Industrial parks and zones are now standard aspects of most of the medium and small cities in the country. Thus far, however, most industrial growth has occurred in larger urban areas, and decentralization policies have had limited impact on rural zones. Substantial implicit subsidies—tax rates, charges for basic services, infrastructure—continue to favor large urban areas despite a history of government commitment to decentralization policies. Moreover, the active role of the state in industrial development continues to encourage location in or near Mexico City where government officials can be contacted easily, licenses acquired, special understandings arranged, and political ties maintained. In consequence of the persistence of such biases toward large urban areas, Rhoda (1983:59–60) argues that the stimulus to relocate industrial capacity to regional centers and small cities should take the form of disincentives to invest in large urban areas, such as through the removal of tax and service subsidies, and disincentives to migrate to urban areas, such as the use of wage restraints and the removal of subsidies for food consumption. In this way, investors will begin to think more positively about the industrial potential of smaller cities and market towns.

There appears to be some potential for addressing the problem of rural underdevelopment through continued emphasis on industrial decentralization. In two of the research communities considered in Chapter 4, the jobs created in small cities by the large industrial sector contributed significantly to incomes in numerous households. Nevertheless, large-scale industry in Mexico continues to be highly capital intensive, and industrial decentralization as it has been pursued is an expensive policy option, often organized around the creation of industrial ports and enclaves that require major investments in infrastructure. The payoff in terms of the number of new permanent jobs created has not been great, although considerable temporary employment has been created in construction. Austerity measures adopted by the government in the early 1980s had a severe impact on these jobs when public and private investment declined significantly.[3] Nevertheless, some room for

3. Some indication of the meaning of the economic crisis for the construction industry is reported by Latin American Monitor's *Annual Report on Mexico*. "The performance of the sector over the last two years pales against the period 1978–81, when the industry recorded growth of more than 10% per annum. Since then, the sector declined 5% in 1982 and 14.3% in 1983 while operating at less than 30% capacity. More than 2m construction workers were made redundant in 1983, while between 40%–50% of the building companies operating in 1982 were bankrupt by the beginning of 1984" (*LAM* 1986:114).

experimentation exists, particularly with regard to subcontracting: "In some countries, certain firms have been able to arrange the making of components in ways that allow the work to be distributed among peasant families working in their homes. Other firms deliberately organize handicraft activities of which they supervise the designs and quality standards, and subsequently effect the marketing of the output. In Mexico, much remains to be tried in both respects" (Yates 1981:251).

Expanding the Maquiladora Program

With regard to industrial decentralization efforts, one promising means of locating large enterprises closer to those rural areas that depend heavily on labor migration is through an expansion of the government's program of in-bond industries. In 1965, the Mexican government initiated the Border Industries Program; in 1971, this program was formalized through facilitating legislation in both Mexico and the United States.[4] The program, created in part to absorb some of the expected labor dislocation caused by the end of the Bracero Program, made it possible for U.S. firms to locate plants in the Mexican border region to produce or assemble goods that would be exported back to the United States without having to pay customs duties on machinery or materials. The United States agreed to tax only the value added on goods produced by subsidiaries of American companies abroad (see especially Fernández 1977; Miller 1984; Seligson and Williams 1981). The U.S. firms could therefore take advantage of low wage rates in Mexico and limited transportation costs to produce goods for the U.S. market.

The Border Industries Program has expanded rapidly in recent years and has been widely credited with having created a large number of jobs. In 1967 there were 57 plants, the so-called maquiladoras, in the border region. In 1980 there were 607 of them, providing 130,000 jobs.

4. The in-bond industries legislation in Mexico allows for 100 percent foreign ownership of plants producing exclusively for export within a 12.5-mile-wide area adjacent to the U.S. border. Firms that compete with Mexican exports are excluded from the program. Imports of raw materials and capital equipment are not subject to customs duties. If all materials and equipment are imported and all products are exported, Mexico assesses no corporate income, sales, or other taxes. Local and state authorities may alter or waive taxes collected locally. The foreign firms are not able to own land in Mexico within 100 kilometers of the border, and they are required to pay the minimum wage. Most of the maquiladoras are engaged in electronic or textile production, but food processing also takes place, and medical supplies, boats, and other products are also assembled (see Fernández 1977:135–137). Currently, major firms with maquila operations in Mexico include Memorex, Samsonite, Magnavox, Litton Industries, Kimberly-Clark, Motorola, Sears Roebuck, Hughes Aircraft, and General Instrument. Often "twin" plants are established, with a U.S. plant doing initial processing and a companion Mexican plant doing assembly. Finishing is done in the U.S. plant.

By 1984, there were 744 plants employing a total of 176,000 people. In this later period, there were reports of labor shortages in the maquiladoras of some border cities, a phenomenon related to the rapid expansion of the program and to the pull of much higher wages in the United States (Miller 1984:6). It was estimated that there would be 800 plants employing 280,000 people by the end of 1986 (*LAM* 1986:108; see also *NYT*, January 10, 1986). The successive devaluations of the peso in the 1980s significantly lowered wage rates and made the maquiladora program more attractive to U.S. firms and even to investors from South Korea, Hong Kong, Singapore, and Taiwan. In 1980 the Border Industries Program generated $1.4 billion in foreign exchange for Mexico and became the second largest foreign exchange earner by 1985 (*LAM* 1986:108; Seligson and Williams 1981:2).

In 1972, the legislation that facilitated the creation and expansion of the maquiladoras in the border region was extended to the entire country, in the expectation that U.S. foreign investment in in-bond industries in the interior of the country would generate the same kind of employment opportunities. After a decade of sluggish growth, these hopes began to be realized only in the mid-1980s, in part as a reaction to labor shortages in the border region. In 1979 there were only fifty-two maquiladoras outside the border region, and none of these was located in a small town or rural area (Seligson and Williams 1981:149). In 1983, 89.3 percent of the plants were in the border zone; by 1986, concentration in this area had declined to 82 percent (*LAM* 1986:109). Recent study of the expansion of the program beyond the border region suggests two principal reasons why the maquiladoras did not initially expand more successfully to the interior of the country. First, local authorities in interior cities, towns, and villages were not assiduous in courting the U.S. firms to interest them in establishing operations in these communities. In contrast, in the early days of the Border Industries Program, authorities in a number of border cities aggressively courted foreign firms, offering attractive tax holidays, investing local resources in supportive infrastructure, and promising to deliver an appropriate and docile labor force for the new industries (see, for example, Whiteford n.d.:24–25). In the interior, however, some reports indicate that local authorities not only lacked such aggressiveness but have at times discouraged foreign investment through personal venality. A second reason cited for the failure of the maquiladoras to expand more rapidly into the interior is the reluctance on the part of foreign management to transfer their activities to areas where they believe they will have difficulty finding or resettling managerial and supervisory personnel and where transportation costs may be higher because of increased distances (see Seligson and Williams 1981:150). These inhibiting factors appear to have been declining in importance because of the increased interest of foreign firms in the program and their need to find a more ready supply of labor.

New interest among foreign investors could be met with more activity by local authorities to attract industry. Presumably, local authorities could be informed more fully of the potential that the industries offer their communities and could be tutored in means of attracting them. Presumably, also, specific incentives to attract managerial and supervisory personnel could be devised and experimented with. In some areas, minimum wage rates that are lower than in the border region could compensate for increased transportation costs.[5] Such measures, if adopted, could prove successful in attracting more in-bond industries into parts of the central plateau region where they would offer alternative sources of income, at least within commuting distance of depressed rural areas. There is also some empirical support for the claim that establishing maquiladoras in the interior could serve to inhibit rural outmigration. Kearney (1984:28) indicates that a soccer ball assembly plant, a maquiladora, in a rural town of Oaxaca not only provided significant employment opportunities but also reduced outmigration from the area. In another study, when workers in the border plants were asked if any equivalent job in their place of origin would encourage them to return home, 47.5 percent of a sample of more than eight hundred workers reported that they would do so (Seligson and Williams 1981:158).

The positive aspects of Mexico's experience with in-bond industries should not be overemphasized, however. The critics of the Border Industries Program are many, and their insights into problematic aspects of the program are important. First, although numerous jobs have been created by the maquiladoras, critics note that these jobs are low paying and the tasks performed by workers both repetitive and intense (see Fernández-Kelly 1983a). In consequence, the turnover rate of employees is very high, with medical and psychological problems often foremost in worker complaints about dissatisfaction with their jobs. Burnout in the young, predominantly female workers appears to occur in as little as two years of initial employment in many plants. The jobs themselves offer few opportunities for upward mobility in the work-

---

5. Mexico has more than sixty-seven minimum-wage zones. In general, wages are lowest away from the border area and outside major urban areas.

place, nor do wage rates significantly increase with seniority. Moreover, critics are disturbed by the dependence on the U.S. economy that is inherent in the expansion of maquiladoras (see Seligson and Williams 1981:2–3). A former high-level official in the Mexican government even stated that "the maquiladoras promote the most demeaning kind of dependency."[6] For these and other reasons, they wonder whether the kinds of employment opportunities created by maquiladora plants are those that Mexico should seek to stimulate within the context of its national development. Others warn that the cost of providing infrastructure to attract the maquiladoras may outweigh the benefits they bring to the local community. Still others indicate the limited nature of backward and forward linkages that often characterize assembly plants. Plans for the expansion of the maquiladora program, which promises considerable capacity to create local employment opportunities, should be tempered with an understanding of these problems. Planners and local authorities should assess the specific advantages of specific types of assembly and production plants before moving ahead with expansionary plans.

Linking Migration to Local Development

Many have argued that, contrary to conventional wisdom, a central condition of rural poverty is not necessarily the absolute scarcity of capital in rural areas but an inability to capture that asset in ways that permit it to be put to productive use (see Von Pischke, Adams, and Donald 1983). According to this view, one key to financing productive ventures in rural development is savings mobilization (see especially Adams 1983; Meyer 1985). In the high-migration areas with which we have been concerned, there is some reason to believe that the mobilization of remittance income would be a viable means of making some capital for local investments locally available and accessible. We have seen, of course, how temporary migration is often undertaken in order to help rural households simply maintain a subsistence or near-subsistence income. In other cases, small amounts of capital are accumulated to improve household welfare through the purchase of consumer durables or housing. In still other cases, surplus income is invested in small enterprise ventures of the household.

6. Interview, November 12, 1984, Center for U.S.-Mexican Studies, University of California, San Diego.

Table 12. Value of migrant remittances through official channels compared with GDP for selected countries, 1985

Country	Remittances ($ million)	GDP ($ million)
Bangladesh	421	16,110
Mali	35	1,100
India	2,291	175,710
Sudan	259	6,930
China	180	265,530
Haiti	98	1,930
Pakistan	2,526	28,240
Sri Lanka	296	5,500
Yemen, PDR	494	900
Yemen, Arab Rep.	897	3,700
Morocco	967	11,850
Philippines	111	32,590
Egypt, Arab Rep.	3,212	30,550
Dominican Rep.	205	4,910
El Salvador	114	3,820
Turkey	1,714	48,820
Tunisia	271	7,240
Colombia	117	34,400
Jordan	1,022	3,450
Syrian Arab Rep.	293	16,370
Portugal	2,075	20,430
Algeria	313	58,180
Greece	775	29,150
Oman	43	8,820

Source: World Bank 1987:206–207, 230–231.

Most studies of temporary labor migration indicate that significant amounts of capital enter rural communities through remittances. Table 12 shows the value of officially reported remittances by migrants from a number of countries.[7] Interestingly, in this form resources flow back into rural areas from more advanced regions, recapitalizing the rural sector rather than decapitalizing it, as is more usual in development experience. It is, of course, impossible to know how much money is

7. Hugo (1977:275) indicates that remittances from labor migration in Indonesia reach 40–45 percent of the weekly income of migrant households remaining in rural areas. According to Berg, in highland Peru, "migrants report that they return with one or two hundred dollars per year spent away, and this is used for the essential costs of setting up a household: paying for a traditional wedding, building a house, and buying livestock, tools, clothing, and school supplies for children" (1985:9). Some comparative guesstimates of monthly international remittances from the Gulf states of the Middle East range from $200–$450 for migrants from North Yemen, to $170 for Pakistanis, to $319–$444 for Indonesians, to $363 for Thai migrants (see Owen 1985:6). For data on Zambian remittances, see Cliffe (1978); on the Philippines, see Griffiths (1979); on Peru, see Berg (1986) and Mallon (1983). On the role of remittances in rural accumulation, see Rempel and Lobdell (1978).

channeled into rural communities through remittances. Much indirect evidence, however, suggests that the amount is substantial. Community-level studies clearly indicate that migrants who seek work in the United States remit considerably more than those who migrate domestically, reflecting the much higher wage scales in the United States. Taylor (1984a:111), for example, found that, in two communities in Michoacán, average remittances from the United States were 34 percent higher than average domestic remittances (see also Cornelius 1976b: 13–14; Roberts 1982:317). In 1977, a migrant in the United States could earn the equivalent of a week's wages in Mexico in one day (Gregory 1986:197). Although we lack recent comparative data, the economic crisis in Mexico, which caused a decline in real wages, and the drastic devaluation of the peso undoubtedly increased the differential between remittances from international and domestic sources.

Most of this money reaches rural locations directly in the form of money orders that arrive through the mail and cash that is carried in the pockets of the migrants when they return. One study attempted to assess the amount of remittances flowing into Mexico from the United States by analyzing records of the Banco Nacional de México for a randomly selected day in 1975 (see Cornelius 1976a:31–35). These data from branch banks throughout the country indicated that 1,117 money orders were processed for a total of $106,704, or $95.53 per money order, a sizable amount for rural Mexico. In the same study, migrant respondents in nine Jalisco communities indicated that they sent money to their families twice a month and that monthly remittances averaged about $200. In-pocket remittances, brought by the migrants when they returned to their community, were reported to vary between $50 and $4,000, with an average of $250–$350 (Cornelius 1976a:31–35). Data presented by Roberts (1982:317) indicate that U.S. remittances averaged $334 per year in one region, and remittances from domestic migration were about $98 annually. In a survey of Hispanic garment workers in Los Angeles, Maram (1980:28–29) found that 37.2 percent of migrants remitted funds at least once a month, and 63.2 percent of the total remitted some income. The average amount remitted monthly was $60.34. Gregory (1986:113) reports that the value of remittances to a poor region of Oaxaca exceeded the value of livestock and agricultural production of the area. Cross and Sandos (1981:118) report that U.S. remittances account for as much as 10 percent of the per capita income of rural inhabitants in the central plateau region. Evidence from the four municípios discussed in the next chapter is also suggestive of considerable amounts of money being sent to migrant families.

Studies of rural communities also indicate that most remittance income goes to the maintenance of subsistence or into housing construction. Once again, economic conditions in Mexico in the 1980s exacerbated this tendency as standards of living fell in rural areas. In two communities in Michoacán, Dinerman (1982:75) discovered that 22.8 percent and 19.7 percent of remittance income was used for family maintenance (see also Cornelius 1976b:14; World Bank 1984:101). In addition, a considerable amount of money is invested in land, with the consequence that in communities heavily involved in migration, land values have risen sharply and social differentiation has increased (Cross and Sandos 1981:45; Reichert 1981:63–64). It appears generally that investment in land is more frequently made to increase household security than to increase income from agriculture (Cross and Sandos 1981:45; Mines 1981:133). The objective of investments in land may differ regionally, however, for Dinerman (1982:74) indicates that land purchases in a relatively fertile area made family vegetable gardening possible and served to increase the dependability of household income. Housing consistently ranks as an important use of migrant remittances, as we will see in the research municípios discussed in Chapter 4. In Dinerman's (1982:75) communities, 24.3 percent and 19.7 percent of migrant income was invested in housing, a larger proportion of total remittances than was invested in any other activity. Indeed, migrant investment in land and housing in Mexico reflects patterns found more widely in Third World countries.[8] Mines (1981:104) found that, over time, investment in assets increased and the proportion of remittances dedicated to subsistence decreased, a trend indicating rising standards of living in migrant households. In the state of Oaxaca, Kearney (1984: 52) found significant amounts of money being spent on civil religious activities such as fiestas and ceremonial functions.

Investments in subsistence have led many to argue that remittance income does not provide a ready source of capital for productive invest-

---

8. A survey of remittance investment in Pakistan indicated that, after subsistence expenditures are accounted for, migrant families tend to (1) repay debts; (2) invest in urban real estate; (3) purchase land; and (4) construct housing. Similarly, in Turkey, some 58 percent of savings from migrant income was invested in housing and land (see World Bank 1984:101). In a survey of the use of remittance income in Bangladesh, rural households used the money to (1) buy land; (2) improve housing; (3) purchase food; (4) repay debts; (5) purchase clothing; (6) invest in agriculture; (7) buy gold and ornaments; (8) invest in business (see Ali et al. 1981:129–136; see also Owen 1985:13–14).

ment. Reichert (1981:63–64), for example, argues that remittance income has allowed migrant households to achieve "higher per capita income and increased consumption. . . , [but] it has not led to the development of the town economy in ways that have stimulated production or created new employment opportunities." Stuart and Kearney are even more pessimistic about the productive use of remittance income, arguing that migration is "a pallative [*sic*] for unemployment, land shortage, lack of credit, and all the other typical conditions of rural stagnation" (1981:36). At the same time, however, these and other analysts admit that the failure to put migrant remittances to more productive use reflects a lack of economic opportunity in the local communities (see Mines 1981:155–156).

The picture is not as bleak as it may appear, however, for there are examples of remittance income that is being put to active and productive use, as cases in Chapter 4 and the following instance demonstrate.

> Before 1967, one of our nine communities was so economically depressed it was losing many inhabitants through permanent emigration. Most of those remaining depended on income earned in the United States. Since 1970, however, the community has grown at a rate of 9.2 percent a year, attracted migrants from surrounding villages and towns, and experienced the greatest boom in its 137-year history. There is, in fact, a significant labor shortage in the community. What happened? In 1967, a migrant who had worked nine years in the United States used the $1,600 he saved to buy two small, hand-operated cloth-weaving machines. Setting up a small factory in his home, he manufactured women's and children's clothing for sale in nearby cities. As the business profited, his neighbors took note. Today the community has about 180 small clothing factories, home-based enterprises that supply clothing to department stores in many cities. Some of the primitive machines have been replaced by sophisticated, motorized machinery—virtually all of it bought with U.S. earnings. Those who continue to work in the United States today are generally middle-aged men who have left families behind to operate the home factories while they earn more investment money to expand their textile production. Few family heads do this, however, because most are able to finance business expansion through locally-generated profits and credit from private banks. One family in four owns a textile factory: the others depend primarily on earnings from jobs in these factories. [Cornelius 1976b:31]

This case is undoubtedly unique, but there are a number of other examples from recent studies of successful ventures financed by remittance income (see Berg 1985; Hugo 1977; Owen 1985). Purchase of an irrigation pump, a sewing machine, a small store, or a tractor by individuals or small groups has permitted some rural inhabitants to expand their incomes significantly (see Cross and Sandos 1981:45; Dinerman 1982:74; Chapter 4). Although investments in housing and land are often dismissed as unproductive, they suggest real opportunities for developing enterprises in construction materials, small livestock production and processing, and vegetable and fruit production where local ecological conditions are favorable.

Given the possibilities for successful economic ventures, more attention needs to be given to ways of capturing surplus remittance and making it available for financing small, locally defined and implemented projects. Other countries have developed schemes for capturing and encouraging remittance income from abroad. In India and Yugoslavia, for instance, it is possible to establish foreign currency accounts; Bangladesh has instituted a program of import permit vouchers that are negotiable at a special exchange rate; and mandatory remittance requirements have been instituted through labor contracts in China, Korea, and the Philippines (World Bank 1984:101). South Korea requires migrant construction workers in the Gulf states to remit 80 percent of their monthly wages (see Owen 1985:6). Such programs do not address the need to capture savings locally, however. Mechanisms that can attract both foreign and domestic remittances to local development activities therefore need to be explored.

One possibility would be the creation of local development banks or cooperatives that have effective savings and loan programs for local inhabitants, including the availability of interest-bearing accounts. These would have the advantage of being locally initiated and locally responsive and could include institutional mechanisms for tapping supportive services and infrastructure from various levels of government. The legal framework for establishing such locally initiated savings and loan associations is problematic, however, because the role of private banks in Mexico has been unclear since the nationalization of the banks in 1982 (see *BLA*, January 2, 1985:1, 3, 5). Cooperatives, which must be legally registered and must conform to a number of organizational requisites established by Mexican law, can nevertheless be a viable organizational form for savings mobilization activities. Mexico already has a fairly extensive network of *cajas populares*, or credit unions, with a large local membership. Market towns and villages often serve as headquarters for small regional savings and loan associations. Attempts to capture remittance income are not without risks, of course. Management problems, corruption, and abuses of power can be expected to plague such activities in the future as they have in the past. Moreover, some ventures using local investment capital are likely to fail through deficiencies in planning, entrepreneurship, training, infrastructure, and marketing. The important point, however, is to begin to think of migrant remittances as a source of considerably more development potential than they have shown in the past.

Mechanisms for capturing rural savings potential need to be closely linked with mechanisms for identifying good investment opportunities. Savings and loan mechanisms could, for instance, be linked to community development corporations to identify and promote viable economic opportunities. Lack of credit continues to be a major impediment to increasing the level of economic development in rural communities (see especially Meyer 1983; see also Page and Steel 1984:24–25). Beyond providing capital for local savings and loan associations, the migratory process can be used to help capitalize development corporations that in turn make loans and grants to local community groups. A unique example of the potential of such activities is the Cooperativa Sin Fronteras, established by the Arizona Farm Workers Union, which began to organize undocumented migrant laborers from Mexico in 1977. In 1978–1979, the Arizona Farm Workers went on strike against citrus growers for whom they worked. The strike was long and bitter, but in the end, growers raised wages and improved conditions as demanded by the union (see Sánchez and Romo 1981). One part of the negotiated settlement was the establishment of a fund to which the growers contributed ten cents per hour of time worked by union members—the amount was later raised to twenty cents per hour. The fund was used to capitalize the Farmworkers Economic Development Corporation (FEDC) in the United States, which was legally registered as a cooperative in Mexico under the name of Cooperativa Sin Fronteras (Reilly 1983). The cooperative, with headquarters in Querétaro, is dedicated to increasing the economic development of rural communities in Mexico and is closely linked to its sister organization, the FEDC, in the United States. The cooperative has a staff of five, and initially the economic development initiatives were centered in the states of San Luis Potosi, Guanajuato, and Querétaro, where most of the migrant workers of the Arizona Farm Workers Union originate.[9]

The cooperative combines a source of funding for local initiatives with an organizational form similar to that of many private development foundations. It makes loans to local community groups, provides

9. Workers also come in significant numbers from Oaxaca, Michoacán, and Guerrero. A small number originate in other states as well (Reilly 1983).

some limited technical assistance and education, and attempts to raise additional funds to expand its activities. Most loans are for small-scale and tangible projects in local communities. The cooperative has been concerned to promote local participation in the development projects, an aspect of its work that is encouraged by the personal relationships between union members and the villages and communities from which they come. Most important, it shows how local initiatives might be financed by taking advantage of the migratory experience.

The Cooperativa Sin Fronteras and the Farmworkers Economic Development Corporation continue to work closely together. In many of the activities, the farm workers act as brokers for the cooperative. Thus, "the farmworkers have recruited midwives from San Antonio to educate midwives in the rural communities, they send improved seeds down, and have acquired machinery for well-drilling."[10] According to observers, decisions continue to be made by the board of the cooperative in Querétaro. The creation of such community development corporations on a binational scale of course requires an organizational base of migratory laborers in the United States such as those organized in the Arizona Farm Workers. Where such a possibility exists, a number of problems may arise from the Mexican government's concerns relative to any organization that receives funds from abroad (Reilly 1983). Nevertheless, the community development corporation model has some potential for generating economic activities in rural communities that can address the employment issue, in addition to promising greater independence and local autonomy from the Mexican state.

Conclusion

A number of potential options for generating employment in or near rural Mexico merit exploration. Some options offer the potential to raise rural incomes but little in the way of direct implications for employment generation, as with agricultural development. Some of these options appear to be more feasible and promising than others, for example small-scale industries. Often the suitability of particular options may be quite site specific, as in the case of agroprocessing. Some employment-generating possibilities are more expensive than others, as is the case with industrial decentralization and infrastructure expan-

10. Interview with Charles Reilly, November 15, 1984, University of California at San Diego.

sion. Some require considerable local initiative, such as in the productive utilization of remittance income. Some are simply not feasible politically, for example the possibility of altering the structure of the ejido.

The choice of employment options to create in rural Mexican communities depends very much on local resources and opportunities. The actual choices for generating economic opportunities will thus tend to be site specific, tailored to the local resource and opportunity base. This characteristic of the activities means that flexibility and adaptability in responding to local opportunities are essential to the success of such endeavors. Reliance on the local resource and opportunity base to generate employment also means that local inhabitants must be active in seeking out and taking advantage of new initiatives, whether these are in farming, small-scale artisanry, service provision, factory work, or other activities. Participation, not passive observance, is essential to ensure that local interests are served and that local opportunities are exploited fully (see Cohen and Uphoff 1977; Esman and Uphoff 1984; Leonard 1982). In addition, my review of income-generating portfolio management by rural households makes it clear that risk is an important characteristic considered by households when they decide how to allocate their labor and investment capital. Risk may therefore be a factor in determining local willingness to participate in new activities that promise economic benefits.

Each of these factors—local resources and opportunities, local interest and initiative, and local attitudes about potential and risk—are evident components of the past, present, and future of four rural municípios described in the next chapter. The constraints faced by these areas are not trivial, and the objective of achieving more self-sustained local development is not going to be achieved easily. In particular, investment capital, initiative, and leadership are critical factors that must be combined to find options for the rural communities. Whether these factors can be made available in any given area may well depend on the attractiveness of the option to migrate.

4

Assessing the Present and Anticipating the Future in Four Rural Areas

The people left behind in the village of Jaripo live on their dreams. They are awaiting the arrival of the son, the grandson, the husband, boyfriend, or father who has gone away to find work.[1] When he returns, they believe, he will come laden with gifts—a television, a stereo, a blender, clothes, and toys. The girls dream of the presents and excitement that will accompany the homecoming, and the boys dream of the day they will be tall and strong enough to leave Jaripo and go with their fathers and brothers to "the other side" or to Mexico City. In the meantime, the girls busy themselves with domestic chores and watch after the young children. When there are no longer enough remittance pesos or dollars to meet daily expenses, the boys work as day laborers for the rich peasants in the area. As they see it, the only escape from this life is migration—a process they can begin when they are fifteen or sixteen years old.

Migration has become a way of life for people in the village of Jaripo because they do not see any economic future for themselves in this region of the state of Michoacán. The soil is impoverished, there is little industry, and the wages paid an agricultural day laborer are attractive only to the destitute. To the people in the village, the future of Jaripo is bleak. There appears to be little foundation for building a more viable local economy, one capable of self-sustained development. How much more sensible to look for work elsewhere and to maintain the regional

1. The description of Jaripo is based on correspondence from M. Celina Robledo, January 30, 1986.

85

tradition of migration that dates to prerevolutionary times, when their grandfathers traveled to agricultural estates in the lowlands to work in the fields.

As we have seen in previous chapters, the story of Jaripo is not unique in Mexico. Over the course of the past several decades, hundreds and thousands of rural communities have gradually lost the capacity to generate employment opportunities for their inhabitants. Rural communities have always been poor, of course, but with their incorporation into larger economic and political systems, they have become increasingly dependent on income generated outside the local area in order to sustain even a subsistence existence within it. Labor migration on a massive scale has been the rural response to a declining natural resource base, increasing population pressure, discriminatory government policies, and deepening maldistribution of wealth. By the 1980s, large numbers of rural communities in the central plateau area of Mexico, like Jaripo, lacked local economies capable of self-sustained development. Most important, their inhabitants no longer believed that they had an economic future in these rural areas. The most telling indexes of these conditions were not simply low agricultural production figures but also extensive labor migration.

This chapter describes and assesses four municípios in the central plateau area in terms of their present conditions and future potential for developing local economies that could generate more employment opportunities for their inhabitants.[2] Each of the research areas manifests a similar set of constraints resulting from its poverty and powerlessness within the Mexican political system, yet each also presents a unique set of conditions that marks its current condition and future potential. In the following pages, these conditions and possibilities will be described on the basis of data collected in the summer and fall of 1985 and in previous years by other scholars. There follows a profile of four municípios in Mexico's central plateau, an area noted for rural population density, rainfall-dependent agriculture, a lengthy incorporation into regional, national, and international economies, and a history of extensive labor migration.

Although the four municípios share histories of government neglect, migration, and constraints set by national and international economic structures, each has a particular condition of poverty and external

2. The criteria for selecting the four municípios are described in the Introduction. The field work was carried out by Merilee Grindle, Elizabeth Meier, and M. Celina Robledo in the summer and fall of 1985.

dependence to overcome if a more viable local economy is to be developed. It will become clear that some of the areas appear to have considerable potential for future development; in others, it is difficult to envision an alternative to their current decline. For some, the most important constraint is the history of external dependence, for others, it is the poverty of the local natural resource base, and for others it is the inhibiting influence of local structures of power and wealth. What can be accomplished in these communities is thus affected by the particular nature of economic underdevelopment in each município as well as by the more general political and economic contexts they share.

In large part, building a more viable local economy depends on the availability of investment capital and incentives to encourage its productive use. Theoretically, of course, such investment capital should come from agriculture, as development theorists have generally argued. In at least two of the research areas, however, it is not easy to understand how capital could be generated from an extremely poor agricultural resource base, even if long-standing policy biases were undone and if infrastructure and technological innovations were introduced. If employment opportunities are to be created in such unproductive environments, sources for investment capital need to be sought elsewhere. In the research communities, there is some evidence to suggest that the productive use of remittance income is one way in which depressed local economies could develop greater demand and generate capital for investment. Thus although extensive labor migration introduces serious distortions into the development of local economies, it also presents some opportunities for locally initiated development. Such local investments are of course only part of the solution to the problems of rural Mexico. Policy changes are also required. The potential uses of remittance income are noted in the community profiles presented in this chapter along with a discussion of other factors that can be introduced to make self-sustained development more possible in each. In Chapter 5, changes in national policies affecting rural areas are considered.

Tepoztlán: Regional Markets and Semiproletarianization

Tepoztlán, a município in the northern part of the state of Morelos, south of Mexico City, is one of the most widely known rural regions in Mexico. It is, in some ways, a famous município, for here, in the 1920s, the anthropologist Robert Redfield carried out the research that led to the development of the concept of a "folk-urban continuum," which has been used to explain the social transformations accompanying development (see Redfield 1930). Later, in the 1940s, Oscar Lewis came to Tepoztlán to pursue the ideas generated by Redfield's work and developed a seminal critique of earlier anthropological theories of social change (see Lewis 1951). In later years, Tepoztlán was studied by Claudio Lomnitz (1978) and was discussed in broader studies by Eric Wolf (1959), John Womack (1968), and Guillermo de la Peña (1981).

Tepoztlán's popularity as a research site is only partially explained by its pleasant climate and beautiful landscape. Its attractiveness to researchers is also based on the município's characteristics, which come as close to typical as is possible in Mexico's highly heterogeneous rural areas. The município concentrates a wide variety of altitudes and climatic zones within its diverse agricultural resource base.[3] Its landholding structures, population growth rates, and age distribution pyramids closely resemble those for Mexico as a whole. The diet, education, and economic activities of its inhabitants are also reproduced in hundreds of other municípios. Tepoztlán's local Indian past continues to survive in conjunction with its nationalized mestizo present, and its economy has become incorporated into regional, national, and international markets. These factors characterize much of the central highlands of Mexico. Finally, the history of the município tells much of the history of central Mexico—the organization of preconquest society, the domination of the colonial period, the disruption of the independence era, the violence of the Revolution of 1910, the agrarian reform of the 1930s, and the gradual incorporation into national economic and political systems in the period after the revolution.

In many ways, then, Tepoztlán is rural Mexico in a nutshell. In particular, its incorporation into the larger economy, the problems of its agricultural sector, and its recently increased dependence on labor migration are representative of a wide range of rural communities in Mexico. Like them, Tepoztlán is a community of notable contrasts. In the fall of 1985, ten-year-old boys played video games in small shops lining the central plaza, bantering in Nahuatl, while their older sisters did the marketing attired in jeans and high heels. In recent years, the vacation houses of wealthy and middle-class residents of Mexico City have filled a landscape still inhabited by people who wear no shoes. Buses leave every fifteen minutes for Cuernavaca and every hour for the capital city, eighty kilometers away. As they wind along the moun-

---

3. Lewis (1951) notes, "Over fifty percent of the total land area of Mexico falls within the range of altitudes found within this single município, that is, from approximately 3,500 to 9,500 feet" (p. xxii).

tainous roads, they pass fields of corn grown for the subsistence needs of poor households. Young people travel to nearby Cuernavaca to work in an industrial park, contributing their earnings to their families on the ejidos and communal lands that surround the town. Migrants to Mexico City and the United States return to celebrate traditional religious holidays in which Aztec and Christian mythology are intermingled. Tepoztlán is full of such contrasts, and in them, the município confronts the major problems of its future. Its economic potential is inextricably linked with Mexico City and Cuernavaca, but these linkages threaten to extinguish its agricultural potential and to condemn its ejidos to increasing impoverishment.

Tepoztlán has never been remote from the main currents of Mexican history or from the economic conditions that characterize large numbers of rural communities. In pre-Columbian times, Tepoztlán was a religious center, a site for pilgrimages and festivals to honor the Aztec god Quetzalcoatl, who is believed to have been born and to have passed his childhood in the area. In 1521, Cortes's troops passed through the town, burning it and forcing its inhabitants to pay tribute to Spain. During the colonial period, the traditional hierarchical Nahuatl social structure was adapted to pay tribute and provide labor service on the Spanish haciendas in the lowlands, in a system extending that imposed by the Aztecs prior to the conquest. Although the município maintained most of its communal lands through the nineteenth century, a large portion of its inhabitants were forced, through lack of water and arable land and unequal access to available land, to migrate to lowland haciendas to work (Lomnitz 1978). In the nineteenth century, considerable class differentiation developed within the município, in large part as a result of unequal access to land and the capacity to control the flow of goods and services into the area and within it (Lomnitz 1978). Cattle raising and trading formed an economic basis for local elites until the outbreak of the Revolution in 1910. Trade relationships included Mexico City and Cuernavaca and extended south throughout the states of Morelos and Guerrero. Such relationships suggest the extensiveness of economic activities at a time when it was possible to enter Tepoztlán only by horse or muleback or on foot. In 1897, the Wells Fargo Company constructed the Mexico-Balsas railroad through the northern fringe of the município. This technological innovation stimulated the incorporation of Tepoztlán into an even broader set of economic and political relationships.

These relationships were destroyed, along with much of daily life in Tepoztlán, by the Revolution of 1910. The município, largely committed to the Zapatista agrarian movement, suffered repeated invasions by competing armies. The years of violence and disruption left a clear mark on the population of the município. In 1910, its reported population was 9,715; by 1921, the population had shrunk to a reported 3,836. The economy of the area was devastated. Villagers who returned to their abandoned homesites in the 1920s were reduced to making charcoal for markets in Mexico City; their fields and livestock had been destroyed by the violence.[4] The population shift, in combination with the agrarian reform of the 1930s, profoundly altered the socioeconomic structure of the município. As in other parts of Morelos, the agrarian reform made it possible to reconsolidate a landholding peasantry that, before the revolution, had been forced increasingly into wage labor. Access to education expanded in the 1920s, and the construction of a corn mill gradually allowed women to become more active in commerce once they had become less tied to the time-consuming job of grinding cornmeal for tortillas. In 1936, a highway was constructed between Tepoztlán and Cuernavaca; local capital was immediately invested in a cooperative bus line. The road initiated a period of transition and growth in the município, offering greater possibilities for petty commerce, educational opportunities, and extensive exposure to urban culture and ideas. It also marked the beginning of a period in which local political leaders were incorporated fully into centralized and national politics (see Lomnitz 1978).

The 1940s brought the beginning of technological change to Tepoztecan agriculture; fertilizer was introduced, and tomatoes and gladiolas were first grown for Mexico City and Cuernavaca markets. With greater commercialization of agriculture, differentiation among types of producers—subsistence and market oriented—became marked because access to capital determined who could acquire land and could afford to transport products to market. The tourist trade developed considerably at this time also, as the first "outsiders" from Mexico City began to come regularly to spend their holidays in the mountains. At about the same time, the first Tepoztecans began migrating to the United States for work.[5] Compared with many other areas, the migratory phenomena came relatively late to the município.

4. Charcoal production requires little capital and technology. In Tepoztlán, a charcoal cooperative was established in the late 1920s and eventually claimed five hundred members before it was disbanded in 1936. The charcoal industry, which continues on an individual basis, has wreaked havoc with the forests of Tepoztlán and has caused considerable internal dissent over its ecological costs (see Lewis 1951:163–165).

5. Lewis (1951:36–37) reports that about fifteen Tepoztecans migrated to the United States to work during World War II, returning to the município with stories of what they had seen and experienced on "the other side."

In the 1950s, a secondary school was built in the município, and commercialized agriculture expanded. The cattle industry declined sharply as a result of a livestock epidemic and the greater returns to using available land for crop production. Electricity came to the community late in the decade after a committee was organized to petition the state government for modernization of the município. Just a short time later the município was fully discovered by the tourist trade, and land was eagerly sought by foreigners and visitors from Mexico City who wished to build vacation homes. Prices for land, services, and consumer goods rose appreciably in the 1960s, reflecting the influence of tourism in the local economy. In the 1960s, another significant event for the economy of the town was the construction of an industrial park in Cuernavaca, about twenty-five kilometers southwest of Tepoztlán, where Nissan, Avon, Ponds, and textile and other manufacturers began production. A continuing process of incorporating Tepoztecans into the wage labor force of Cuernavaca and into national labor organizations dates to this period. In fact, this incorporation and the expansion of tourist-related services had proceeded to such an extent by the latter part of the 1960s that a labor shortage developed in local agriculture. Commercial farmers, still producing tomatoes and flowers for Mexico City and other markets, began to import laborers from the states of Guerrero and Oaxaca for periods of peak labor demand.

By the mid-1980s, Tepoztlán had become a relatively urbanized municipal center. More than half its total population of 21,768 people lived in the center in 1980. At that time, there were fourteen primary schools, an intermediate school, and a secondary school. Eighty-seven percent of the population over fifteen years of age was classified as literate, an indication of the relatively urbanized character of the município. The municipal center was a busy place. More than eighty commercial establishments offered an impressive array of goods and services. A weekly market filled the central plaza with more than a hundred stalls selling a wide variety of foods and consumer goods. The transport, services, and construction industries grew rapidly to reflect the tourist industry, and manufacturing expanded in the period after 1960. By the 1980s, 11 percent of the município's population was engaged in service occupations, 7 percent was in manufacturing pursuits, 5 percent was in construction, and 5 percent was in commerce. Small artisan production of items such as clothes and jewelry also developed. In part, these activities were encouraged by tourism. Local women, for example, were successful in organizing and sustaining a cooperative to produce cotton and wool shawls, shirts, dresses, and blouses. The cooperative began as a private venture in 1979 with an investment of some eight thousand dollars. Initially, three seamstresses were employed, but by the end of 1980, thirty were working for the business venture. In 1983, a number of employees, with the encouragement of the owners, purchased the business and formed a cooperative, which continued to grow, in part because of the stimulus of the local tourist industry that made direct sales a feasible way to address marketing problems. In the mid-1980s, the cooperative of fifteen members employed forty seamstresses and supplied markets in Mexico City and the United States.

In spite of these opportunities, the people of Tepoztlán claim there is no work and little future in the town. Increasingly, they set their eyes on Cuernavaca, Mexico City, or the United States. Tourism, they claim, does not offer substantial rewards, and there is no opportunity in agriculture to earn a living wage. In fact, in the 1980 census, 25 percent of the economically active population in the munícipio claimed no specific occupation, suggesting a high level of unemployment or under-employment. Public works kept a number of people employed in the 1970s and early 1980s, when potable water was brought to the entire town and roads were paved. The economic crisis of the 1980s brought such activities to a near halt, however, and severely constrained the availability of investment capital that might have been used to develop the construction materials and furniture industries for which raw materials are available locally. According to local inhabitants, the economic problems of the 1980s were so great that wages from Cuernavaca and remittances from the United States that used to be set aside for small investments were increasingly used to meet subsistence needs. The inflationary impact of tourism was cited frequently as a factor in forcing households to diversify their income.

In spite of such pessimism on the part of local people, the municipal center did offer some alternatives for the future of Tepoztecans in local manufacturing, services, and commerce. Moreover, the center had an infrastructure of roads and services and a suitable location for investment in small-scale industrial activities linked to markets in Cuernavaca and Mexico City. It also had a substantial array of public institutions and services—schools, clinics, CONASUPO stores, banks, and government offices. In the seven villages flanking the municipal center, however, a different perspective was evident. In the 1980s, these villages continued to be characterized by an economy based on subsistence agriculture supplemented by short-term wage labor in agriculture. In these communities, the 44 percent of Tepoztlán's economically active population that claims agriculture as its occupation depended primarily on the production of crops such as corn, beans, and nopal. Poor soils and the virtual absence of irrigation mean that most households produced little surplus. Abundant rain falls during the summer and fall months, but during the rest of the year, precipitation is insufficient for agriculture. As a result, there is only one crop cycle during the year. Droughts occur with some frequency in the region, but the principal problem is overabundance of water during some periods and its complete absence during others.

Access to land is a major issue in the communities, along with the lack of resources to invest in agriculture. Of the total land available in Tepoztlán, only 16 percent is suitable for agriculture. In the villages, many ejidatarios sharecrop their own land. The very poorest supplement family income through the sale of wood. Although 25,906 hectares of a total of 27,833 are officially registered as ejido and communal lands, there has emerged over time a notable concentration of control over communal tenure lands. Moreover, of the 1,917 hectares held in private hands, almost half is in parcels of five hectares or more. In addition to the problem of land concentration, there is simply not enough arable land to meet the needs of the agrarian population. The scarcity of land—and the value placed upon land ownership—was noted in the 1940s by Oscar Lewis (see Lewis 1951:118, 125–128). Forty years later, ownership of and access to land continued to be a source of conflict in Tepoztlán's rural areas. The development potential in Tepoztecan agriculture is extremely limited except perhaps in the sense of encouraging the production of locally marketable crops for regional self-sufficiency that might help counteract the impact of tourist-related inflation.

Many young men from these seven villages commute to the industrial park in Cuernavaca each day and supply the income to make it possible for households to remain in their villages. As economic conditions worsened in the 1980s, temporary migration to the United States increased. In one village, for example, 32 workers went to the United States and Canada in 1984; in 1985, 130 left for these destinations. Most of the international migrants stay away for three to eight months at a time, and very few choose to remain in the north. Generally, migrants from Tepoztlán have been able to go legally to the United States through an institutionalized relationship with labor contractors who help arrange papers and who have even organized short courses on how to behave in the United States. Most migrants send remittances to their families prior to returning. In the past, the remittances were often accompanied by orders to purchase construction materials so that they would be ready when the migrant returned and was able to improve or add to the family's housing. Others have used remittances to pay for their children's education or to invest in small businesses. Increasingly in the 1980s, however, remittances were needed to ensure mere subsistence.

San Juan Tlacontenco, a village of 1,600 inhabitants—some 180 families—demonstrates the problems of Tepoztecan agriculture. At seventy-two hundred feet above sea level, the village is the highest in the município. Although the land is officially registered as communal, private control over it has become extensive. Two economic strata characterize the village: a small number of rich peasants who control large amounts of land and who have access to tractors and trucks, and the poor peasants who have only very small plots of land and no access to machinery. Rich peasants also own the two general provision stores in the community, the only visible economic activity beyond agriculture. The road into the village often becomes impassable during the rainy season, a difficult problem because of the economic dependence of the village on Cuernavaca and Tepoztlán. The village is noted for the production of nopal, which is marketed by the women in the two market centers.[6] Nopal production, however, is primarily an activity of the poorest peasants, who lack access to the technical assistance and credit necessary to make it a more remunerative crop. The lands of richer peasants are used primarily to produce forage crops. A few peasants grow fruits such as peaches and pears, and flowers, but generally the quality of these products is not high. In addition, a few households raise sheep, although this does not form a significant part of the local economy. There is no irrigation to expand the number of crop cycles or the variety of crops produced.

Clearly, then, infrastructure development in terms of better roads and irrigation systems and assistance to producers in terms of credit and extension would provide a more effective basis for the local agricultural economy, even though the lack of arable land would continue to be a limiting factor on productive potential. San Juan's experience with prior development efforts has not been encouraging, however. In one attempt to provide water to the community, a well was dug, but the available machinery was not powerful enough to pump the water to the surface, and the project was abandoned. Communal efforts continue to supply the village with water, but the conditions of the roads and poor organization have impeded effective resolution of this problem. Other de-

---

6. Nopal is a form of cactus that is cooked and consumed by low-income people in Mexico. The plant has potential in the manufacture of medicinal and cosmetic products.

velopment projects have made little contribution to improving conditions in San Juan. The village has a medical service one day a week that is provided by the national government, but its operation is highly erratic. The Ministry of Education and the Ministry of Agriculture have technical assistance projects in the village, but they are directed to only a few individuals, usually the better-off members of the community. A more positive project, focused on the intensive production of nopal, has been carried out by the Center for Experimentation for Technical Development, a private organization supported by the Organization of American States. When the project was initiated in the early 1980s, each family enrolled in the project received two thousand plants, necessary pesticides, and the technical assistance needed to modernize production. With appropriate credit, production of nopal in the village was raised some 80 percent over the course of several years. The project envisioned the possibility of local processing to produce shampoo, cosmetics, and medicinal products. San Juan is also a place where small food-processing activities with locally grown fruits might be introduced. These plans could not become reality, however, unless the problem of investment capital were addressed, because the only economic surplus in the village has been extracted elsewhere, through migratory labor flows. Migration, perhaps because it is a recently initiated phenomenon in the community, has not had much impact in terms of investment in income-generating activities in the community.

San Andrés de la Cal, about ten kilometers from the municipal center of Tepoztlán, is also a village characterized by communal landholdings and few economic opportunities. The land, at forty-nine hundred feet above sea level, is rocky and hilly; moreover, there is not enough of it to go around. Much of the village's communal effort in the past has been to acquire more land through the agrarian reform. This history, dating to the 1920s and 1930s, is a familiar one in Mexico—petitions, conflict with neighboring communities over rights to land, bureaucratic and political delays, and population growth mitigating the potential of new lands. In the 1980s, San Andrés had a population of thirty-four hundred people—some 400 families—a few of whom continued to speak Nahuatl. Agricultural production was characterized by the commercial production of tomatoes and the subsistence production of corn and beans. A larger community than San Juan, San Andrés was also more stratified economically. The most advantaged stratum was made up of people with adequate land and access to machinery and green revolution technology. A second stratum was made up of people with small plots who employed traditional technology. At the bottom of the economic pyramid was the largest group, those who had no land and who depended on wage labor. Even those with sufficient land, however, were not able to make it produce effectively. At least one person in each family was involved in wage labor or in informal sector activities such as street vending. In most cases, individuals would go to surrounding towns to work as day laborers in agriculture. In the mid-1980s, seven men from the village had secured permanent employment in the industrial park in Cuernavaca, where they earned slightly more than the minimum wage and in addition received social security, loans, and paid vacations. Migration to the United States and Canada, never a strong tradition in the community, began to accelerate noticeably in the late 1970s and was becoming common by the mid-1980s. Remittances, reported to amount to two hundred to five hundred dollars a month, were generally spent on improved housing and education when they were not fully needed for household subsistence.

Although San Andrés appears to offer few opportunities for employment, some of its residents have invested their migratory earnings to generate viable economic activities for their families. One good example of the use of resources flowing back into the community from migration is the bakery business started by Don Pablo. In the early 1980s, he built an oven with the surplus of remittances sent by his sons from the United States. His investment was worth approximately $800 in 1985, and he supplied some one hundred families with bread. In that year, Don Pablo estimated that his monthly income is approximately $387 dollars. Nicéfero, twenty-three years old, migrated to the United States in the early 1980s. With his savings from that experience, he purchased a used Volkswagen for $1,350 and worked as a taxi driver in Tepoztlán, earning about $64 a week. He owned his own house and paid day laborers to work his land. Other families have also used migration to improve their income. Ana worked her family's plot of land and tended the children while her husband, Agustín, went to Canada to work for eight months. When he returned, he brought enough money to expand the house and purchase a television and a sewing machine. Ana began to make quilts and aprons for sale in Tepoztlán. Although her earnings were not great, she believed that they accounted for a third of the family income. In another case, Doña Rosita managed to build up a small business without migratory remittances. For seven years, she raised pigs until she had earned a profit of $1,650. She used this sum to purchase an inventory of clothing to sell in the community; in a single year she was able to double the size of her original inventory.

In spite of these economic activities, however, San Andrés remains marginalized from greater economic dynamism because of its poor agricultural resource base and because of factors such as the expense of transportation. At a time when the local minimum agricultural wage was $2.85 a day, the cost of traveling one way to Cuernavaca was $0.81 and passage to Tepoztlán cost $0.40, a clear disincentive to families wishing to educate their children beyond the local primary school or to work beyond the boundaries of their own community. High transport costs also made it difficult to market agricultural products. These deficiencies indicate how the development of infrastructure might increase the economic potential of San Andrés. In a pattern similar to that in San Juan, little had been accomplished in the past through public programs to improve local economic conditions. A technical assistance program in agriculture of the Ministry of Education, for example, accomplished little to improve methods of production in the community. An adult education program initiated by the same ministry lacked a teacher with sufficient interest to keep it alive. The Ministry of Agriculture claimed to have a project in the community, but no one in the village was aware of it. Another government agency organized a sewing cooperative that involved twelve women; they subsequently abandoned the project because of its top-down management style.

A more positive experience was a chicken farm supported by the Organization of American States. The project began in the early 1980s with an investment of $24,500. Its purpose was to provide cheaper food for the community and to develop a local basis for protecting inhabitants from the inflated prices of the regional tourist economy. Within three months of its initiation, the farm became self-supporting. A second phase of the project involved the purchases of ten cows for dairy production. Again, the project was successful in generating income for the nine individuals it employed. An obvious need for further expansion of this project was the local production of animal feed for the livestock. Local raw materials were available for this process, and plans for such an activity were discussed and assessed as feasible by community residents. Investment capital was lacking, however, and promoters of the idea had little sense of where they might secure it.

The economic problems of the villages of Tepoztlán are exacerbated by the poor quality of available land, lack of water, the small size of holdings for poor peasants, increasing landlessness, inadequate infrastructure, and the inflation in the local economy caused by tourism. On the other hand, Tepoztlán is close enough to Cuernavaca to benefit from the availability of jobs and markets for locally produced items. In some cases, migration has had a positive impact on the local economy as migrants have returned to use their remittances to create durable sources of income for themselves and their families. Clearly the future of Tepoztlán will involve an ever closer nexus between the município and the cities of Cuernavaca and Mexico City. This situation creates problems for the peasants who can no longer live from their land and who must increasingly interact in a local economy that is distorted by tourism. For them, retaining a hold on the land will require increased official prices for the products they produce, a greater effort to provide irrigation and other infrastructure, and projects to encourage the productive investment of local capital. In addition, attracting investment capital for labor-intensive industrial pursuits could bring jobs and increased incomes to the villagers of Tepoztlán. In the absence of such innovations, the rural poor are likely to be drawn more fully into the orbit of repeated migration in order to make ends meet.

Jaral del Progreso: Commercial Agriculture and Class Structures

The município of Jaral del Progreso is the richest of the four communities considered in this study. Its richness is most evident in the late summer and early fall, when the extensive fields surrounding the municipal center are green with the approaching harvest. Production on the flat valley land is irrigated, mechanized, and technically advanced; it is hard to escape the overall impression that Jaral is a prosperous and modern rural region. During these months of lush growth, the municipal center is almost deserted, and the tree-lined central plaza is quiet except for the presence of a few elderly men. Everyone else, it seems, is at work in the fields. Jaral is very different in appearance from Tepoztlán but not surprisingly so, given its location. The município is set deep within the Bajío, the richest and most technologically advanced agricultural area of central Mexico (see Roberts 1984, 1982; Tejera Gaona 1982).[7] It is in this area that the green revolution and irrigation facilities introduced in the 1950s and 1960s combined to produce a modern commercial agriculture whose products are shipped north to the United States and are used domestically to supplement a rapidly growing livestock sector.

In the 1980s, feed grains, notably sorghum, had expanded to such an

7. The region of which Jaral is a part has been extensively studied by Kenneth Roberts. He uses longitudinal data to trace changes in production and costs of production for different-sized agricultural units in the region and explores the nature of increasing stratification that has accompanied agricultural modernization.

extent in the Bajío that one analyst referred to their production as the second green revolution (see DeWalt 1985). Agricultural change and innovation were of course not new phenomena in the region. Since the time of the conquest, commercial agriculture had characterized production in the Bajío, an area centered in the states of Guanajuato and Michoacán. Haciendas made their appearance in the region in the late sixteenth century and rapidly began to serve a regional market that developed around the silver-mining industry. The eighteenth and nineteenth centuries were also characterized by rapid agricultural expansion that differed considerably from the hacienda-based development in other regions (Brading 1977:29). Because the region was not heavily populated with indigenous communities, access to a large and stable work force was always a problem for the haciendas. As a result, hacendados introduced tenancy arrangements to ensure that labor would be available to them (Brading 1977:30). When the population of the region increased significantly in the nineteenth century, and competition over access to land grew, heightened labor demands were placed on tenants and day laborers alike. Nevertheless, during this period, many of the large estates fell into financial crisis and were broken up into smaller "ranchos." With time, a class of relatively prosperous smallholders developed in the Bajío, and the Revolution of 1910 provided them with the opportunity to consolidate some control over land (Brading 1977). Subsequently the agrarian reform carried out in the 1930s and 1940s further differentiated the Bajío from most other regions in that some of the rich valley land was distributed as ejido grants, along with the more normal pattern of distributing dry and mountainous land. The Bajío was significantly affected by the introduction of improved varieties of wheat in the 1950s, when the first green revolution transformed the area. Then, in the 1970s, sorghum rapidly replaced corn as the principal summer cycle crop. This change was so great that it was an important factor in the national decline in basic crop production (see Roberts 1984).

Noticeable changes in the area have occurred since the 1950s. In 1954, when the region was intensively studied by a group of investigators from the Centro de Investigaciones Agrarias, corn and beans were grown on nearly 90 percent of the land during the summer months. Winter wheat was the dominant crop in the second cycle (Roberts 1984:4). In fact, the Bajío was the most important wheat region in the country in the 1950s. According to the study team, the use of green revolution inputs such as chemical fertilizers, improved seeds, and pesticides was limited. Although much of the crop area was commercialized and irrigated, marked differences among ejidos and private producers were not evident (see Roberts 1984:4). This situation changed in subsequent years with the rapid agricultural modernization of the zone. In 1974 a team from the Centro de Investigaciones Agrarias restudied the area and found radical increases in yields in corn and wheat and the widespread substitution of corn by sorghum, to the extent that sorghum came to account for 41 percent of the land planted in four major grains. In addition, greatly increased use of machinery and the addition of costly inputs in production were noted by the researchers (see Roberts 1984). The area had clearly been affected by agricultural modernization, and as a result, yields and overall production levels had increased dramatically.

These changes did not benefit all farmers equally. The 1974 research team documented the emergence of a significant economic differential among types of farmers and between irrigated and unirrigated land in terms of returns to producers (see Roberts 1984:19–27, 33). In part as a result of these changes, the agricultural wage labor force increased much more rapidly than the overall agricultural labor force (Roberts 1984:28). Increasingly, poorer farm households in the region received significant portions of their income from wage labor on the large irrigated farms. In addition, income from off-farm labor was increasingly generated in the local cities of Celaya and Salamanca and in the United States. The local economy became more modern and diversified but also more stratified between the "haves" and the "have nots."

As is typical of the Bajío, the basis of the economy in Jaral del Progreso is agriculture, but it is not the subsistence or semisubsistence farming that characterizes much of rural Mexico. Instead Jaral boasts a dynamic agricultural sector that is firmly embedded in national and international markets. Sorghum, white corn, sunflower, wheat, vegetables, and fruit are grown with modern techniques even on ejido lands. Irrigation is extensive. Two agricultural cycles mark each year, one in which wheat and barley are grown and one in which sorghum and corn are produced. Where vegetables are grown, three crops are produced each year. Sorghum is sold in the nearby towns of Querétaro, Irapuato, and Salamanca to companies such as Purina, and vegetables are shipped north to the United States. Wheat is sent to Mexico City, Toluca, and Puebla. Moreover, in addition to its rich agricultural base, Jaral is located near the industrial centers of Celaya, Salamanca, and León, which provide jobs for the município's residents.

As a result of its rich agricultural base and its integration into national and international markets, Jaral del Progreso offers considerable potential for economic development. Government policies for agricultural and industrial development have favored rather than discriminated against the area, and it is in the lead in the utilization of modern agricultural techniques. Nevertheless, the area is marked by extensive labor migration to the United States, a characteristic found frequently in poorer rural regions. In Jaral, few families have not sent someone to "the other side." In spite of the evident richness of the area and the access that ejidos have to irrigation and modern technology, the level of poverty of a large portion of its population is similar to that found in much more depressed zones. These conditions have obviously not developed because of a poor local resource base, as in the case of Tepoztlán and Unión de San Antonio. Instead, the source of the extensive evidence of poverty in Jaral must be sought in the inequitable distribution of economic and political power in the region, structures that grew directly from the modernization of its agricultural economy.

Jaral lies about 250 kilometers north of Mexico City and is reached easily by the main north-south road that connects the capital with the cities of Querétaro, Celaya, Irapuato, León, Aguascalientes, Zacatecas, Chihuahua, and eventually Ciudad Juárez and El Paso, Texas. The town of Jaral del Progreso was founded in 1590 and was declared a município in 1863. In 1980, Jaral had a population of 15,646 in the municipal center and forty-four surrounding villages. It had become one of the best-irrigated municípios in the state of Guanajuato. Agricultural land was about evenly divided between ejidos and private property. As I have indicated, ejidos had some access to land and irrigation of good quality. In fact, the basic characteristics of landholding that differentiated poor farmers from commercial producers was not whether the land was communally or privately held but whether or not it was irrigated. This difference became increasingly evident in the 1960s and 1970s. Just as in the Bajío more generally, agriculture in Jaral was pursued primarily on the basis of traditional techniques until the 1950s. The predominant crops were corn and wheat. Fertilizers and pesticides were not used. Production fluctuated considerably from one year to the next in response to the availability of rainfall. In the 1950s, the major rivers of the area, which form part of the Río Lerma system, were dammed, and irrigation became widely available. Then, in the late 1950s, sorghum and sunflower were introduced and were widely adopted by farmers in the município. In the next twenty-five years, the economy of Jaral was transformed as part of the overall transformation of the Bajío.

Along with agricultural modernization, migration became an institutionalized aspect of the regional economy. Increasingly, households in

Jaral sent family members to search for jobs off the farm. Migratory flows correspond directly to the need for labor in agriculture. In the peak sowing and harvest times (May-June and November-December), extensive demand for labor provides considerable income-generating opportunities in Jaral. In the months before and after these periods, however, peasants in the area generally have work only one or two days a week. The solution to this household-level problem of seasonal underemployment has been temporary migration to Morelos and Veracruz to cut cane, to Colima to work as masons, and to the United States to work in agriculture. Permanent jobs are found in nearby towns of Celaya and Salamanca. Income generated in industrial work contributes significantly to sustaining many rural families. Although the município is close to these and other industrial centers, however, there is little industry in Jaral. In fact, a small factory producing balanced animal feed and two small family businesses producing cheese and sweets are the only evident activities of this type. Local inhabitants argue that the answer to Jaral's problems of poverty and underemployment is to introduce more agroprocessing industries in the município, especially for balanced feed and vegetables, and industries based on agricultural machinery and implements, such as machinery repairs and the production of spare parts. Few believe such changes are possible, however. The reason, they assert forcefully, is the embedded class structure of Jaral.

The município is clearly divided between those who have adequate land and irrigation and those who have only small plots of land and no irrigation. In addition, a growing sector of the population is landless. The first group reaps the benefits of the green revolution in the Bajío; the latter two groups are excluded from it. Thus a small and wealthy class of farmers has emerged from the introduction of highly commercialized agriculture in the region. These wealthy farmers live in Jaral, many of them in chateauxlike houses on the main road into the municipal center. This economically advantaged class owns its own land and rents or sharecrops a considerable portion of the ejido land. Although the available statistics do not adequately describe conditions surrounding access to land in the region, it is widely acknowledged in Jaral that considerable concentration of control occurred in the 1960s, 1970s, and 1980s. Large landowners may control three hundred to four hundred hectares, often by accumulating land in the hands of different family members, whereas poor farmers are fortunate to have access to three or four. Wealthy farmers often have contracts with brokerage firms in the United States for the production of fruits and vegetables. Access to irrigation and to national and international markets for local crops continued to be principal factors separating the advantaged from the disadvantaged, however (Roberts 1982:309).

Creating jobs through regional industrialization efforts is one obvious way of attempting to meet the needs of poorer sectors of the population in Jaral. Infrastructure, markets, and even investment capital are available locally to support such an initiative. An inhibiting factor in the introduction of more job-creating opportunities, however, and especially full-time jobs, is the opposition of the wealthy farmers to any initiatives that might affect the availability of a large and reliable agricultural labor force. Residents of the ejidos and villages surrounding Jaral recount numerous examples of local elites sacrificing the economic advancement of the município to their own need for a large seasonal work force. The influence of these elites in making local opportunities available, for example, is clear in looking at the experience of three villages in the município, Zempoala, Victoria de Cortázar, and Santiago Capitiro.

Zempoala is an ejido some fifteen kilometers from the municipal center. It is a poor ejido that has had electricity only since 1975 and potable drinking water only since 1982. Of the eighty families composing the ejido, sixty-five have rights to the land; the rest are landless and likely to be barred from future access to land because of insurmountable political obstacles to increasing the size of the ejido. Class consciousness is acute in Zempoala. Local residents are both cynical and bitter about the resources of land and influence controlled by *los ricos*. In discussions with community inhabitants, Jaral was often called "the great hacienda" of the rich—those who controlled access to land, markets, finance, and political power. One resident characterized the local economy as belonging to the "clans" of rich families who hoarded all available resources.

Class sentiment in Zempoala developed to such a marked state in part because of an unsuccessful effort of the ejido to acquire more land. In 1980, after years of petitioning, the ejido was granted an extension of 104 hectares by the government. Titles were then transferred to the ejidatarios. Plans were made to plant the new lands, located in the richest agricultural zone of the município. When the planting was done, the ejidatarios camped out in the fields to protect their crops from the anger of the wealthy farmers, who claimed the land was theirs. The Zempoalans, however, were no match for the tractors that these individuals had driven onto the land to destroy the crops. The ejidatarios retreated. Added to the bitterness of their experience is the disdain of the ejidatarios for the political authorities at the state and national level who permitted and then supported this abuse of their rights. Zempoalans now believe they have no capacity ever to take possession of land again. Their response has been to give up the fight for the land and to look to other alternatives for the survival of the ejido. The land that was legally granted to the ejido remains in the hands of los ricos.

For most people in Zempoala, the alternative to working their own land is work as seasonal agricultural laborers on the lands of the wealthy farmers for $2.60 a day. Women and children join the agricultural work force in peak periods but are paid less. In the late afternoon they can be seen walking back to the ejido from the fields of los ricos. For those who continue to work ejido land, sorghum, corn, and wheat are the only crops that can be grown; producing the more remunerative vegetables requires access to the markets and credit controlled by los ricos. There is little economic activity other than agriculture in Zempoala. Four small stores attached to dwellings sell staples and an astonishing variety of junk food and soft drinks. One person raises ducks and pigs for sale. No local residents have cars or trucks to provide taxi or marketing services, and there is no evidence of cottage industries. Inhabitants are certain that installing a factory "of whatever kind" will solve their economic problems. They are equally certain that "los ricos will not permit it because it would cost them their agricultural labor force." As a result, they are resigned to the idea that nothing can be done except to migrate to "the other side" for work. Indeed, a very large proportion of the male population has been to the United States. For them, it is well worth the risk and the six hundred to eight hundred dollars needed to get there. "You can see the difference it makes in how people live," commented one resident.

Victoria de Cortázar is only seven kilometers from Jaral, although it takes thirty-five minutes to make the trip because of the poor condition of the road. Victoria is a village with a more differentiated economy than that of Zempoala. It boasts telephone, mail, light, and drainage services as well as a good number of paved streets, nine locally owned taxis, two buses, and a library. The road into the village is bounded by extensive fields sown with sorghum and interspersed with irrigation canals. The settlement was an extensive hacienda situated between two branches of the Lerma River until 1922. In that year, the area was declared to be a town. In the 1930s, land was distributed as part of the agrarian reform. Currently, the settlement contains both private and ejido lands, and 180 ejidatarios each control between one and four hectares. Crops produced include wheat, sorghum, corn, barley, tomatoes, beans, alfalfa, hay, and vegetables as well as melon and a variety of tree fruits. The level of technological development on the larger land-holdings is impressive. Tomatoes, onions, cauliflower, and broccoli are grown for companies such as Birdseye, Del Monte, and Green Giant, all of which have processing plants in nearby cities.

Before sorghum and other crops were widely adopted, Victoria de Cortázar produced corn and sugarcane. The empty shells of several sugar mills dotted throughout the landscape attest to the importance of sugar in the local economy in the past. The sorghum boom of the 1960s and 1970s rescued the town from almost certain economic collapse when sugar ceased to be viable economically. By the 1980s, it was clear that its rich agricultural economy had brought numerous benefits to Victoria. Nevertheless, the extensive mechanized production of sorghum, wheat, and barley severely reduced the amount of employment available. As a result, many inhabitants have turned to migration—to Mexico City, Morelos, Celaya, Querétaro, and "the other side." The youth of the town talk knowledgeably about San Bernardino, Santa Ana, downtown (Los Angeles), and their experiences of being arrested by the border patrol. These same young people have been active in organizing community betterment projects and in contributing to local improvement activities. Despite extensive migration, Victoria de Cortázar is a place to which they want to return.

In addition to migratory labor, there are three artisan workshops or small factories in Victoria. One produces baby blankets and quilts, another produces slacks and shirts, and the third produces canvas packs for schoolchildren. The firm that produces blankets and quilts has a considerable investment in machinery. The owner lives in Celaya and employs twelve people in Victoria. At times and depending on demand, the factory works all three shifts. For the employees, the work is attractive because the pay is considerably more than in agricultural labor. According to one worker, for instance, "I earn $2.60–$3.20 a day in agriculture and here I earn $26 a week. What is more important to me is that there is no work in the countryside because of the machinery that's now being used. That's why I'm in this factory." There are also some cottage industries in Victoria; for example, one family produces open-work embroidery, another makes clay vases and ornaments, and another manufactures small tortilla baskets. These activities use only family labor, and the household views them as supplementary sources of income to agricultural pursuits.

As with the residents of Zempoala, the people in Victoria de Cortázar believe that the solution to the employment problem in their community is industrialization. In spite of the evidence of self-initiated small industries in the village, they look to the government to provide them with the appropriate factories. Their demands are clear. "If there is sorghum, why don't we have mills for it? Why don't we have a canning facility if we have vegetables? The government should make plans on the basis of what we have and introduce appropriate industries. The government of this state has promised us these things; let's see if it comes through." In the meantime, the young people anticipate migrating to Chicago, California, Texas, Florida, and New Mexico. Many go every year, and some find it convenient, because of the risks involved in crossing the border, to remain for extended periods. In their attitudes about the future, they are much like rural inhabitants in countless other communities throughout the central plateau area of Mexico who see little that they can do to promote the development of their towns. At the same time, however, they return again and again to their homes, unwilling to give up completely on their future.

Santiago Capitiro, like Zempoala, is an ejido belonging to the municípío of Jaral. It is situated about ten kilometers from Victoria de Cortázar. The village has light and water services but no drainage, pavement, or mail service. It boasts a central plaza with a concrete communal pool where women congregate to do the family washing. Santiago Capitiro is a poor ejido with insufficient land to sustain its population. Like Victoria de Cortázar, it was a hacienda before the agrarian reform reached the region; in 1936, it was declared an ejido, and ninety-three ejidatarios each received an average of four hectares. Thus a large portion of its fifty-one hundred residents is landless. Like all other communities in Jaral del Progreso, the village has an economy based on agriculture. Corn, wheat, beans, sorghum, and vegetables are grown; corn and sorghum predominate. Small-scale pig raising has developed as a specialty of the community, in large part because there is so little land available. In addition, a small factory produces slacks and employs about thirty-five women in the ejido. Some fifteen small stores provide the residents with staples and agricultural supplies, and a market once a week brings considerable commerce to the village from surrounding areas, but these activities are insufficient for the economic needs of the inhabitants of Santiago. As in other rural settlements in Jaral, young people generate a sizable portion of household income, migrating and sending remittances to their families. Some go to Celaya, and some go to Mexico City, but most go north to the United States. In the 1980s, local residents had little belief that their village could develop a more dynamic economy; their world was constrained by los ricos who controlled adjacent lands and by the option to migrate. Their views compound the dilemma of development in the município.

Jaral del Progreso is a rich município. Nevertheless, its very richness has created obstacles to its development: concentration of landholding, seasonal unemployment, displacement of labor by machines, dependence on migration, and above all, a power structure that makes it difficult for local inhabitants to envision other alternatives for their own future. The solutions to such problems require organization, leadership, and political support. The current structure of power in the município and the attitudes of resignation that it has generated among the poor suggest that these characteristics will be difficult to acquire. In such a situation, local efforts are likely to be spurred most effectively by the introduction of some outside initiative to increase perceptions of the options that are available.

Unión de San Antonio: Livestock and Migration

The contrast between the agricultural lushness of Jaral and the aridity of Unión de San Antonio is striking. Although the second município is only about 180 kilometers northwest of Jaral, its agricultural resource base is clearly distinct. The fields and hills surrounding Unión are sparsely covered with semidesertlike flora. There is much less evidence of agricultural production; where cultivated fields are to be found, they are farmed far less intensively than those of Jaral. Cattle graze in groups of five or ten on sparse pastures throughout the landscape. Few people can be seen working in the fields, nor can they be found congregating in the central plaza of Unión, an attractive square of green leafy trees and white benches surrounded by buildings in the Spanish colonial style. The quiet of the town and its surrounding agricultural lands is broken only by the sounds of the construction work in evidence everywhere.

Unión, a pleasant town to visit, appears surprisingly prosperous. Just as the poverty of Jaral was difficult to understand, the source of the evident prosperity of Unión is not easy to identify. Even in outlying villages and ranchos, television antennas are visible from the roofs of brightly painted stucco houses, many of which have uncharacteristic picture windows facing the street.[8] Elaborate metalwork is often used as

---

8. A similar observation was made by Ina Dinerman (1982) in a discussion of migratory patterns in Michoacán.

grills on windows or as ornaments on front porches. Large pickup trucks and jeeps careen through quiet streets. Many of the men in the town sport pistols on their belts. The young women of the area are turned out in stylish jeans and high-heeled shoes. The parochial church is large and well tended. The municipal offices are new, attractive, and spacious. Why such evidence of prosperity, especially when local inhabitants lament the poor state of the local economy and the lack of opportunities it offers? This developmental anomaly of the município can best be understood through an assessment of local economic and social relations.

Unión de San Antonio has a long history dating to 1563, when the first families arrived and the first of the region's large land grants was made. In subsequent years, haciendas established by the settlers tended to be of medium size. Modest landholdings were necessary because there was not a large indigenous population that could be exploited as an agricultural labor force. Over the years, inheritance patterns and population pressure reduced the size of the early landholdings until they were relatively small. Moreover, the haciendas were always more oriented toward livestocking than to crop production largely as a result of the poor quality of local soils, lack of water, and the absence of a large labor force (Craig 1983:25).[9] The nineteenth century brought considerable economic expansion to the region; the "golden years" of the local elite in nearby Lagos de Moreno correspond to the presidency of Porfirio Díaz (1875–1910) (Craig 1983:30). The Revolution of 1910 had a considerable impact on the município, reducing the size of its population to thirty-six people, according to one local resident. Especially in the years 1910–1914, large numbers of people left for the city of León, where they felt protected from the violence of contending armies. The period most etched in local history, however, followed the revolution, in the late 1920s, when open violence flared between adherents of conservative Catholicism and the new secular state (see Craig 1983; Meyer 1976). The Cristero rebellion lasted into the 1930s in this region and left enduring enmity in local relationships among those who consider the Catholic church and its officials to be bastions of moral and political righteousness and those who see it as an instrument of class interests and repression. A significant flow of migration to the United States began in the late 1920s as people attempted to escape the violence and

9. The region that includes Unión de San Antonio has been extensively studied by Ann L. Craig (1983) and Wayne Cornelius (1976a). Craig provides important historical information on the município, and Cornelius analyzes the migratory behavior of its inhabitants. Both individuals have generously made their rich field notes available to me.

economic turmoil of the Cristero Rebellion. In the agrarian reform that was carried out in the 1930s in Unión, those who rejected the idea of petitioning for recognition as ejidatarios were most closely associated with the hacienda structure and local Catholic clergy (see Craig 1983: 30).

The inhabitants of Unión de San Antonio are proud of their Spanish roots and the attractive Spanish flavor of the constructions in the blocks surrounding the town plaza. The parochial church dates to 1801 and is a favored landmark of local residents. The município, in the eastern part of the state of Jalisco, lies just south of the small city of Lagos de Moreno. It is about eight hours by car from the center of Mexico City. Its population of 13,201 inhabitants (1980) is widely dispersed in 104 small "ranchos," each consisting of only a few families; this settlement pattern makes service with water, electricity, sanitation, and health facilities difficult and costly. The same problems may affect the education of the inhabitants. In 1985, the município had seventy-six primary and two secondary schools; in spite of considerable investment in education, 30 percent of the inhabitants were listed as illiterate. According to the 1980 census, 49.6 percent of the economically active people in the município claimed agriculture and cattle ranching as principal occupations. Construction, manufacturing, and services, the most dynamic sectors of the local economy, increased significantly in the 1970s and 1980s, whereas commerce remained relatively stagnant. At the same time, workers without specified occupations grew in number, reflecting the increase of landlessness on the ejidos and the diversified nature of individual activity.

The explanation of the apparent contradictions of Unión—prosperity and lack of visible economic pursuits, television antennas and lack of education—is not to be found in agriculture. The average size of an ejido plot is seven hectares, too small to ensure subsistence on such arid land, and not very promising for cattle ranching in an area in which one hectare of grazing land is required at a minimum for every head of cattle. Private landholders control plots that average twenty-six hectares. The implication is clearly that ejidatarios suffer from a shortage of land, especially given its poor quality and the extensive erosion that affects agricultural potential. In fact, however, in recent years farmers have left about 15 percent of their land uncultivated, often renting it out to cattle ranchers. Renting and sharecropping have been particularly evident on ejido plots. Corn, the principal crop, is produced using traditional techniques and little or no financing; almost 90 percent of it is grown for home consumption. Beans, also consumed by producers, are the other principal crop of the region. Since 1980, sorghum has been introduced, but it has yet to displace more than a small fraction of the corn produced.[10] Sorghum is grown with more modern production techniques than corn and beans, but few inhabitants can make ends meet effectively on the basis of agriculture even with more modern technology. The lack of water is a critical constraint on agriculture, as noted in the following description.

> There are marked seasonal changes in the appearance of the land, changes which accentuate the vital importance of water. In the winter, cloud-free skies, bright sunshine, and pleasant temperatures contrast with a dusty, parched, brown landscape. Particularly in the driest parts of the region [of Lagos de Moreno], by March or April all of the reserve water in earthen dams has evaporated or been drained off for cattle, home use, and some irrigation. . . . Cacti stand out as welcome greenery. It is difficult to spot the adobe houses at a distance, so easily do they blend into the landscape. But by August, one would never recognize the hills and fields as part of the same landscape. With even scant rainfall, grass and wildflowers sprout . . . and . . . the landscape is bathed in hues of green, purple, and blue. [Craig 1983:21]

Although a poor natural resource base constrains agriculture, cattle ranching provides a more positive picture of the economic potential of Unión. Ranchers depend primarily on natural pastures rather than balanced feed for their animals. Gradually, agriculture has been giving way to an expansion of ranching. Most of the cattle have traditionally belonged to small private producers, with only 7.7 percent of cattle lands belonging to the ejidos. Given the uncertainty in crop production, many ejidatarios prefer to rent their lands to the more prosperous ranchers. Most of the livestock is in dairy cows, although with bad harvests the production of milk tends to go down and that of beef production to accelerate. Local industry, focused on the production of milk, cheese, and other dairy products, reflects the economic importance of the livestock sector. Milk production is relatively technified and the influence in local production of the establishment of large processing plants owned by Nestlé and Danesa has been considerable. Nestlé, for example, provides technical assistance, forage, and some credit at low prices to its contracted producers. These services have increased the supply and quality of milk produced.

Although large transnational corporations dominate the production

---

10. In 1985 only 335 hectares in the municípío were planted in sorghum.

of dairy goods in the region, a number of small independent producers of cream and cheese supply the local market. In 1975 there were thirteen such establishments in Unión de San Antonio, employing forty-three people. In 1985 there were twenty small establishments employing some sixty people. The largest of the small cream and cheese producers is Los Caporales, established in 1960 when its founder decided to produce cheese and then cream with his five dairy cows. Using family labor, he began production in his home; profits were invested in expanding his herd until he had twenty cows. In the mid-1960s, he employed his four sons and six additional workers. Production techniques were never fully modernized, however, because the owner was unable to acquire credit from the local bank. His own cows provide about 75 percent of the raw material for his business, and local milk producers supply the rest. In addition to Los Caporales, a sizable factory was established in 1983 to produce milk and whey; it employed forty people in the mid-1980s. The factory also provided technical assistance to its contracted suppliers of milk, who tended to be the larger producers. In spite of the permanent employment it offers, turnover is high in the factory. Young people often use their wages to pay the coyotes and to meet other costs of being launched into the migratory network to the United States.

The dairy industry somewhat explains the apparent prosperity of Unión de San Antonio, but it does not give the entire picture. There are household industries such as sewing and knitting and small commercial outlets for food, clothing, shoes, and household and agricultural goods. A number of taxis and buses service the community. In one village, women pass the day knitting and crocheting baby clothes, which they sell to intermediaries, who in turn market them in specialty shops in Lagos de Moreno, León, and Guadalajara. Each of these contributes a small amount of vitality to the local economy. The real solution to the economic puzzle of Unión, however, is migration, perhaps the most important activity of the município. Migrants go to both domestic and international destinations, although the United States is the most frequently mentioned objective, and the volume is sensitive to conditions in the local economy. Domestic migrants go to the cities of Guadalajara, León, San Francisco de León, and Mexico City to work as brickmasons and wage laborers. Migration to the United States is well institutionalized, with a tradition dating to the 1880s. Most of the migrants find work in agriculture, although the number found in service and manufacturing has been increasing with time. The stimulus to this migratory pattern is clear to local inhabitants: a poor natural resource base, an underdeveloped agricultural sector, and a dairy industry that employs few people, either on the farms or in the food-processing sector. In addition, the dispersed nature of the population constrains the development of jobs in the service and commercial sectors.

In contrast to other areas, Unión has not experienced the pressure on the land that has stimulated migration in other areas. The dynamic in Unión is the abandonment of agriculture because of its poor economic payoffs. In fact, the quality of land in the region is so poor that the agrarian struggle for access to land has been muted in recent years. The reason is that, by the mid-1980s, few local inhabitants believed that their economic problems could be resolved through access to land. In some ways, then, Unión is a rural town that has given up on agriculture. This has not always been the case. In the 1920s and 1930s, there was considerable conflict as agraristas and conservatives battled for control over land and access to power (see Craig 1983). According to surviving agraristas—those who fought for the land—the worst land in the region was organized into ejidos, so that even those who won access to it lost in the end. Cattle ranching continued to be the most viable economic activity, but it provided few jobs.

Not surprisingly, then, migration has replaced the land in terms of local perceptions of economic opportunities. According to one observer of local events, "The tale of this place is simple. There are few conflicts over acquiring more land. All the land is assigned to ejidos and small private holdings and no one has much hope of acquiring more land or making it produce more. . . . Because of this, all the young people go to 'the other side' to work; besides that, migration is a tradition. But those who go, always return." This statement is not borne out in fact, however, as the município's population was smaller in 1980 than in 1970 by more than one thousand inhabitants. The extensive nature of migration has also encouraged the expansion of livestocking, an economic solution that is pursued at the cost of jobs in the local community. An old peasant and former agrarista summed up his perception of the situation simply.

> We don't want to lose our lands because this is what we old folks fought for, at least to have a roof over our heads and a few beans to eat. But now the land does not even produce this little bit and when it doesn't rain, there is nothing. Today the young people aren't as interested in defending the land because they go outside, earn more and then bring back money for fiestas or to improve their houses, but not to invest in businesses. . . . Even the women want to go and earn money outside. They aren't like the women used to be.

Remittances from migration have been useful in initiating some local businesses. One cheese-manufacturing firm has been in operation for eighty years and was started by a returned migrant early in the century. Currently, the small factory produces three hundred kilograms of cheese and forty to fifty kilograms of cream each week. The owner, who inherited it from his grandfather, employs three workers. Another cheese producer learned his trade in an established factory and then left to go to the United States. In five years he saved enough money in agricultural labor to return and invest in his own small factory of cheese and cream. Slowly, he invested earnings in machinery and livestock. He now employs four workers and sells his products, using his own truck, in nearby cities in the states of Guanajuato and Jalisco. The history of another producer is similar; he used money he earned in the United States to begin a family business currently employing four brothers. The problem that each of these businesses face is marketing. National and international markets are controlled by the transnational businesses. Small entrepreneurs must depend on local and regional markets by working through people they know in nearby cities and towns who are willing to sell their products. According to local wisdom, the most effective thing a young person can do to "make it" economically is to leave his land in the care of others, migrate to the United States, work hard, and save a little capital to bring back to Unión to invest in housing, children's educations, and perhaps a small business. Those who are successful in doing this reflect the local supposition that, on the whole, migrants are those who most want to better themselves.

Remittances explain a large part of the prosperity of Unión and also indicate some of the positive benefits that can be garnered for a community through migration—the investment in productive activities for which there is local or regional or national demand and which can free individuals from the need to migrate repeatedly. Still, the migration process takes its toll and, for most households, has become a recurrent necessity. The local priest keeps a registry of "absent sons" who are considered residents of the town even though their annual visits are of short duration.[11] Although residents of Unión profess to love and to be proud of their community, they also tend to see their economic prospects and even their future elsewhere. One young woman reported, "I make cheese. That's what my mother taught me. I have a boyfriend but he is braceriando. . . . I'm not frightened by his going, everyone does it. I want to go too, and earn more. There's just not enough for us to make

11. Field notes of Ann L. Craig and Wayne Cornelius.

ends meet here. . . . If I go, I have friends there who will help me learn English. Nothing bad will happen to me there."

The village of Tlacuitapa is, of all the communities in Unión, the one most affected by dependence on migration. This village of forty-two hundred people is populated primarily by women and children ten months of every year. In December and January, it takes on the appearance of a more normal community except that there is little evidence of economic activity other than the constant crocheting of the women. In fact, the village has become noted for beautiful hand crochet work in baby clothes. Virtually all women spend a part of their day making the tiny garments for sale in nearby towns. In Tlacuitapa, a member of the local elite supplies the thread and yarn and buys the completed garments at a preestablished price from the women. Most are able to make two or three items of baby clothing a day; once the cost of the thread and yarn has been deducted from their earnings by the local buyer, they realize a dollar to a dollar and a half in profit. The women are well aware of the exploitation they suffer at the hands of the local buyer but have done little to alter the situation. Some of the women have discussed the possibility of forming a cooperative to purchase supplies and market their products, but there appears to be little enthusiasm for initiating such an activity.

Agriculture in Tlacuitapa is based on private landholdings averaging twenty to thirty hectares each and ejido lands averaging about six hectares, belonging to some 146 ejidatarios. Until 1970, the community had no electricity and no adequate road. There is no industry other than the household production of baby clothes. One local resident explained this lack of economic activity in terms of migration. "Why put in factories if everyone wants to leave?"[12] Most agricultural activity involves raising pigs for home consumption and cattle for dairying. Much of the ejido land has been rented out to cattle ranchers. Agricultural experts in the region believe that there is some potential for commercializing crops such as sorghum, alfalfa, and forages in the village but that to do so will require construction of numerous wells and provision of credit for mechanization. In the meantime, they complain that producers suffer because there are too many animals raised on the land, a condition that results in low production and increased erosion. Local residents in turn complain about municipal officials' failure to show interest in aiding their village. A frequently voiced concern is that the town of Unión has progressed at the cost of Tlacuitapa and other

12. Field notes of Ann L. Craig and Wayne Cornelius.

villages. Here again, however, local inhabitants seem little inclined to alter this situation and look to migration as an easier solution to their economic problems. In fact, for assistance with local development projects, the town looks neither to its local officials, who have been engaged in factional and even violent political struggles for decades, or to officials at the municipal, state, or national level.[13] Instead, they look to the association of former residents of Tlacuitapa that lives in Mexico City and that meets regularly to reaffirm regional ties and to discuss ways of assisting the Jaliscan "homeland."[14] The Mexico City migrants are permanent residents in the capital, but they maintain strong ties to Tlacuitapa and are concerned about its future. It is hard to escape the impression that they are more concerned about its future than the residents who continue to live in the village but whose economic base is elsewhere.

The ejido of San Antonio de la Garza graphically illustrates the dilemmas of the município's development. It is a small village of eight hundred inhabitants about six kilometers from Unión and Tlacuitapa. Its agricultural activity in corn and beans is carried out with minimal technological input and has been gradually giving way to cattle ranching through rental of ejido lands. Yet the village does not appear to be poor. Although the only enterprises on the main road through the town are three large cantinas, the village is full of multicolored stucco houses with tall television antennas protruding from zinc and tile roofs. In 1985, the village employed a number of local residents in digging up the main street to put in potable water pipes, a community improvement for which remittance contributions paid in part. In fact, of course, the main economic activity of San Antonio, and the one that explains its prosperity, is migration.

An official of the local bank in Unión de San Antonio aptly described the dilemma of the development of the entire município.

> Conditions for agriculture here are difficult; the land is poor, there isn't enough water, and rainfall is scarce. You could probably promote agriculture here with large capital investments in infrastructure. But large private investors are not interested in this because of the long period before they realize returns on their investments. Investment in livestocking is, in contrast, much less risky and has a relatively short return on investment. . . .

13. A local cacique, or rural boss, who had controlled the town for twenty-three years, was assassinated in the early 1970s by a rival faction. Local politics continues to revolve around the competition of local caciques and their followers. These groups are linked to rival leaders in Lagos de Moreno and Jalisco politics.
14. Field notes from Ann L. Craig and Wayne Cornelius.

From whatever point of view, however, livestocking employs few workers and because of this peasants and their children seek work elsewhere, principally in the United States. This issue of migration is not new, though. It has become a custom of the people and they take it as a natural and unquestioned phenomenon. . . . Those who are most likely to emigrate are ejidatarios because they have the poorest economic possibilities here. . . . They rent out their lands to ranchers, which is their only alternative for earning money here. . . . Workshops and cooperatives that would employ primarily women would help somewhat. . . , but this would not solve the migration problem. The answer is in food processing and manufacturing. . . , but then the question is, Who's interested in this? In any event, this would be a long term solution. . . . In my view, in these municípios with a strong tradition of migration, there is now no solution. The rhetoric of the politicians insists there is and they make lots of empty promises, but the people are not organized to demand alternatives.

Remedying the situation is not an impossible task. Some investments could help Unión become a more economically viable município. It is well located to make use of regional, national, and international markets, and it has an experienced and hard-working local population. Industrial development in this region could make sense economically. Even local improvements, such as road building, have offered some an alternative to migration. In the early 1970s, migration from one village in the município was significantly diminished for two years when a labor intensive local road was constructed.[15] Such solutions, however, require active local commitment to attracting investment and investors willing to consider the region's potential. Neither of these conditions is likely to emerge spontaneously; their development indicates the need for local initiative and an ability to think beyond the migration syndrome. A local analyst of the situation is pessimistic about the potential of Unión to work collectively to find solutions to its problems. "Here, individualism predominates and everyone fights to maintain the status quo: the rich in order to enrich themselves more, the middle class to stay as they are, and the poor in order not to die of hunger. What is common to all of them is that they want to be better off than they are; what is distinct is that many of them have to leave in order to pursue this dream." Migration has brought much to Unión in terms of remittance income and development projects and there is potential for it to contribute even more effectively to the economy of the region. At the same time, the possibility of migration helps explain why the residents of Unión do

15. Field notes from Ann L. Craig and Wayne Cornelius.

not organize economically or politically to find solutions to their economic problems.

Villamar: A Future on the "Other Side"

Although some positive benefits of migration are evident in the experience of Unión de San Antonio, it is difficult to be optimistic about the future of Villamar because of the extent to which the migratory cycle has diminished the viability of the local economy. Set in the northern part of the state of Michoacán, about four hundred kilometers from Mexico City, this município demonstrates that migration can become a complete substitute for local economic activity. The widespread experience of going to "the other side" has created a situation of virtual dependence on the migratory cycle for the income needs of a large portion of the population. For much of the year, in fact, Villamar is populated by women, children, and the elderly. Even commercial activity is stagnant because local entrepreneurs see little point in investing in an area in which economic activity is minimal and erratic. Generally, migrants leave Villamar in February. From that time until they return in November, the municipal center is virtually a ghost town. The migrants return faithfully each year in order to spend the holidays with their families. At this time, it is common for them to use their earnings to buy an animal, to improve their houses, and to consume a considerable amount in local cantinas. The year is thus punctuated by the rhythm of the migratory cycle, not by the needs of local agriculture as one would expect in a rural area.[16]

The roots of Villamar's dependence on migration lie in its economic history. The município has never enjoyed a vigorous local agricultural economy. Its inhabitants have, since the seventeenth century, depended on agricultural wage labor. In that period, large haciendas expanded rapidly, claiming extensive lands for cattle ranching and sugar production. Small villages were swallowed up by the expansion of the haciendas or, like Villamar, have lost so much of their agricultural land that it became common for residents to hire themselves out as day laborers on the haciendas. By the early nineteenth century, the incorporation of the region into the hacienda economy was complete. An independent peasantry had disappeared. Either all local communities

---

16. A village in Villamar has been extensively studied by Omar Fonseca and Lilia Moreno (n.d.).

had been physically incorporated into the haciendas and their inhabitants had become resident laborers, or the villages had been so deprived of alternative sources of income that there was little choice but to hire out as day laborers on the haciendas. A community's success in maintaining its physical, if not economic, independence from the haciendas continued to be an important issue in the region long after the hacienda structure was destroyed by the agrarian reform of the 1930s. Residents of some villages continued to be proud of the fact that their forebears had never been resident laborers and had, as they interpreted it, maintained some independence of spirit. This distinction was also important in an economic sense, for it tended to separate communities that wished to benefit from the agrarian reform (the physically independent settlements) and those that resisted it (the incorporated settlements).[17]

The hacienda-centered past has not been forgotten in Villamar. The largest of the large landholdings was La Guaracha, established in the eighteenth century. By the turn of the twentieth century, it covered thirty-five thousand hectares. The work on the hacienda is still remembered by local inhabitants who were wage laborers for the landowners until 1936, when the agrarian reform ended its life.

> Our situation on the hacienda was very rough. We worked from sun up to sun down. . . . The salary they gave us was three pesos a week and of this the company store took one and a half pesos, leaving us with only one and a half. . . . Many people, seeing that the salary was not enough, stole things. Since the thefts were always discovered, those who robbed were put in prison or lashed by the guards of the hacienda. And the thefts were always discovered because the priest, who was allied with the overseers, would tell them the confessions people made.

Villages in the Villamar region lived on the wage labor provided by the haciendas like La Guaracha until well into the twentieth century. The work was regimented by a hierarchical system of bosses; common laborers worked in gangs of fifteen to twenty people on single tasks, and their activities were closely monitored. There were even individuals whose job it was to verify that those claiming illness were indeed too sick to leave their beds. The wage laborers supplied their own hand implements, and the hacienda was responsible for supplying animals, carts, and plows. Villagers who were not incorporated into the hacienda as a resident work force had to provide for their own subsistence during

---

17. A very similar history is recounted in IMISAC (1982). The agrarian reform reached the area between 1928 and 1932, when Lázaro Cárdenas was governor of Michoacán.

periods when there was no demand for labor in the corn, bean, or cane fields. Their very independence came at considerable cost in added economic responsibility, but the villages were not all equally affected by the incorporation into the hacienda labor force, particularly those that were able to maintain some access to land. In the nineteenth century, for example, a few people in the village of Jaripo were able to rent land in sharecropping arrangements to supplement daily subsistence needs that were not met through wage labor.

Some peasants from the region also began to travel far afield in search of employment in this period. Thus a migratory tradition to the lowlands was well institutionalized by the turn of the century; by 1910, workers were traveling regularly to the "hot lands" to work on the sugar haciendas and to acquire higher wages than was possible in their highland zone. Higher wages continued to be the lure that stimulated the search for work. It became part of local economic thinking that one could always do better elsewhere. By the 1920s, an annual migratory flow to the United States increased over time to become the major destination for those looking for employment. From the beginning, migration to the United States was also seen as a way to earn high wages relative to those paid for locally available work. Moreover, residents continued to go to the "hot lands" to work in the rice harvest each year in the search for higher wages. Land tenure relations, a modest natural resource base, and the attraction of high wages elsewhere continue to make Villamar a place that its inhabitants leave; they return with great loyalty, but only for short periods and only in order to relax and maintain their family ties.

In 1970, 23,806 inhabitants were counted in Villamar; in 1980, 20,757 were counted. In the same decade, the number of houses decreased from 4,056 to 3,566. The decline in the population, according to municipal officials, is due to migration. Migration, they claim, is a symptom of socioeconomic problems in the município that have few solutions. In fact, they underscore their pessimism by noting the extent to which permanent outmigration is occurring in addition to the normal pattern of temporary labor migration. With further inspection, however, the socioeconomic problems of Villamar seem less pressing than those of other areas. The agricultural potential, for example, is greater than in Tepoztlán and Unión de San Antonio. The local economy is so stagnant apparently because of the extent to which migration has become a normal economic activity for large numbers of households residing there.

Villamar is not significantly poorer than other rural communities in central Mexico. Its land resources, although far from sufficient or well endowed by nature, are more productive than those of Unión de San Antonio. Between thirty-two and forty inches of rain fall annually, and the climate, at four thousand feet above sea level, is warm and pleasant. The município is very near the agricultural fields of Zamora, which have become a center for the labor-intensive production and processing of fruits and vegetables that supply winter markets in the United States. In Sahuyo, another nearby town where people in Villamar often go to shop, there is a significant local shoe industry. Other centers within commuting distance have developed textile, chemical, milk, and processed animal feed industries. Villamar has developed no industries of a comparable nature. Instead, it is a município with a labor force that is largely rural, and it has little in the way of local commercial or other activities.

In part, these factors are reflected in basic data about the population. A significant proportion of Villamar's inhabitants live in dispersed localities of fewer than twenty-five hundred people. In all, 66.8 percent of the population was considered rural in 1980; this population lived in villages, only about half of which had potable water and electricity. Moreover, about 60 percent of the population claimed agriculture as its principal activity, even though it is commonly acknowledged that most of the agricultural labor takes place in California, not Villamar. Services, manufacturing, commerce, construction, and transport occupied some 27 percent of the population, and 13 percent claimed no specific occupational category in 1980. The município also has a large dependent population; 45.5 percent was under fifteen years of age in 1980 and another 6.2 percent was over sixty-five. In the age group between fifteen and sixty-five, women predominate.

About a third of the available land in Villamar is dedicated to cropping, another third is used for dairy cattle, and the remainder is in forests. Ejidatarios and small farmers grow corn, wheat, beans, saffron, and garbanzos on small plots, some of which are irrigated. Only recently have vegetables and sorghum been introduced, and few farmers have adopted the production of these crops. The size of ejido plots varies between two and five hectares, although smallholders may have access to as much as twenty to fifty hectares. Even on the larger plots, however, few modern techniques are employed. The terrain is not suitable for extensive mechanization, and individual plots are subdivided. The majority of those with small plots produce only for self-sufficiency. Perhaps even more common than owner-operator production, however, is the increasing importance of rental and sharecropping arrangements.

Renting has been one means by which livestocking has expanded in Villamar. The quality of dairy herds, however, is generally low; extensive rather than intensive production techniques predominate. Milk is sold to intermediaries in nearby towns of Jiquilpan and Zamora or to the government's marketing agency, CONASUPO. According to local producers, uncertainty about land tenure rights, extensive use of renting arrangements, lack of institutional support for credit and technical assistance, and an inefficient and costly marketing system keep them from modernizing their livestock management. Nevertheless, a few producers have imported improved breeds and have begun to introduce balanced feed. Most households also keep a pig or two as a form of family savings. The Ministry of Agriculture does not maintain statistics at the municipal level for production in Villamar, so unimportant is it; rather, figures are maintained for an aggregate of several municípios in the region. The extensive livestocking and renting arrangements, in addition to subsistence farming, suggest some underlying constraints on developing a more dynamic agricultural sector. The modest quality of the agricultural resource base explains part of the limited investment in agriculture, but the principal constraint on its development is the shortage of available labor.

In contrast to Unión de San Antonio, where there is little work in agriculture, there are significant opportunities for agricultural labor in Villamar. Larger landholders complain that they cannot acquire enough seasonal labor to make further investment in agriculture a feasible alternative. The daily wage of $3.86, of course, is hardly an encouragement to those seeking work. This amount is barely enough for household subsistence and provides no opportunities for savings or bettering one's position, according to local residents who choose not to work locally. Thus those who wish to employ agricultural labor in Villamar are hard pressed to find a traditional labor force. As a result, in the 1970s and 1980s, they turned increasingly to female labor. Even then, however, agricultural lands are often left unplanted because of the lack of labor. Cattle ranching, which requires little labor, continues to expand onto agricultural lands.

Migration is frequently discussed in the villages of Villamar. According to local residents, half or more of the population of small hamlets may be absent for the greater part of each year. The low local salaries in agriculture and the lack of alternatives are universally blamed for the migration. According to the reputed cacique, or rural boss, of one village:

> The majority of men go to the United States not so much because of hunger but because of ambition. Many want to improve themselves, seeing how a salary of 1,000 to 1,200 pesos a day doesn't even cover their needs and how others have spread the word about migration. It is no wonder that the town is practically empty. This situation has created problems for me to the extent that I have had to hire women as agricultural laborers. Men are so scarce that I generally hire between 15 or 20 women.

The problems in agriculture have generated the strong migratory tradition of Villamar; migration in turn creates conditions in which there is little incentive to invest in agriculture.

The same problem affects interest in investing in nonagricultural activities. As with the other municípios, local residents see the solution to their economic problems in local industrial development such as agroprocessing. Just as in the other municípios, however, few have any ideas about how such industries might be started. Municipal resources are "raquítico," according to officials, and they see little promise in state or national government activities. If they do not have the resources locally, they see little possibility of interesting other levels of government in their plight. Officials are similarly at a loss to think of private investors who might be interested in the area. For them, the more advanced development of other municípios in the same part of Michoacán only serves to highlight the backwardness and stagnation of their own communities.

Added to these problems of envisioning an alternative future for the município are a series of rancorous local political tensions in which local public officials align themselves with competing caciques, who make it difficult to achieve even modest municipal improvements. One local official, who excoriated her political rivals for not doing anything to help the community, explained that improvements were possible only because of the fondness of the migrants for their home region. "As I told you," she commented, "the municipal president isn't at all interested in the town. For example, when I wanted to get a few streets cobblestoned, I asked for help from him and he only gave me 38,000 pesos. If it had not been for the émigrés, I never would have been able to do anything. I sent word for help to the United States. In that country, some people put themselves in charge of collecting money and they sent it to me. For the cobblestoning, they sent me a million and a half pesos." In fact, this same official had turned repeatedly to the migrants for assistance. The tapping of migrant interest in their communities is similar to the experience of settlements in Unión de San Antonio.

Remittance income has therefore brought some improvement to community and individual life in Villamar. On the whole, however, the depressed condition of the local economy acts as a disincentive for productive investment of remittances. This point is demonstrated in the experiences of a number of local migrants. Miguel, for example, left for the United States in 1980 because he felt there was nothing to do in his town and because he was not willing to work for the low salaries offered in agriculture locally. In the United States he earned enough money to buy a house and a few animals. Each year he would spend six or seven months in the United States, always returning at the end of the calendar year to his home village of Jaripo. He claimed that he never had any difficulty in crossing the border; his destination was always a community in California where he had relatives who assisted him in finding a job. Throughout his migratory career, he remained a loyal son of Jaripo with a home and family in his native community. He was proud of the fact that he frequently contributed to collections among his village mates in the United States to be used for improvements in his town. In an interview in 1985, he indicated a strong desire for his local community to have a factory of some kind so he would not have to continue to migrate. Even if such a factory were to exist, however, he expressed the opinion that he would probably not earn enough money to dissuade him from making his annual trip. For Miguel, the strong tradition of migration in Jaripo had become a way of life.

Jesús told a similar story. He first went to the United States in 1953. Before beginning life as a migrant, he had a piece of land he sharecropped, but the income he derived from it was so unreliable that he preferred to try to migrate to California to work in agriculture. While he was migrating each year, he continued to work his land in Jaripo. In 1971, he bought a house in the village and abandoned his agricultural pursuits. Instead, he invested in a small ice cream business. In 1985, when he was interviewed, he claimed he had never made a good living from his investment. He blamed the massive migration of the region for the lack of opportunities in commerce in Villamar. Because there was little security or growth in local business, merchants were unable to expand and offer residents the array of goods they wanted. As a result, Jesús claimed, most people would go to surrounding towns in other municípios when they had a purchase to make and merchants in Jaripo were left with little business. Like others, he observed that, each year, more Jaripeños were persuaded to try migration and fewer returned each year. Rubén, another former migrant, echoed the pessimism of Jesús. Between 1965 and 1976, he migrated annually to Chicago, where he worked as a mechanic. During these years, he accumulated enough money to buy some land, a few animals, and the largest store in Villamar. By the 1980s, however, he was reducing the size of his store because there was so little business. Once again, the culprit was seen to be migration. During most of the year, there was so little economic activity in the town that it did not make sense for him to continue to try to eke out a living from commerce.

The problems of Villamar are perhaps the most difficult of the four municípios to address. Not only do local inhabitants see little possibility for an alternative future, they also seem to lack any desire to see a more dynamic economy develop in the area. In many ways, they have given up, and the possibility of continuing to migrate appears easy compared with the imagination and energy necessary to break the cycle of dependence. There are alternatives that can be developed in Villamar: agriculture, if given adequate incentives, could be modernized; wage labor in nearby towns could be expanded; processing and other industries might be located in the municipal center. The município is not isolated from interaction with regional and national markets that might be used to advantage, nor is the educational or social development of the area an impediment to more economic development. Given current perspectives, however, it is not likely that alternatives such as those mentioned above will be pursued. Like Unión de San Antonio, this is a situation in which local leadership and local government might make a difference. In Villamar, however, local leadership appears too involved in internecine conflict to think about a brighter future for the region, and local residents are too used to the migrant syndrome to think that any other alternatives might be preferable to current reality.

Conclusion

Bringing change and lessened economic dependence to Tepoztlán, Jaral del Progreso, Unión de San Antonio, or Villamar is no small task. Poverty, discriminatory policies, class structures, migratory traditions, poor natural resources, and political conflicts are among the factors that inhibit the economic development of these communities. Currently, most of them lack investment capital, local leadership, and even the aspiration to free their local communities of dependence on temporary migration to keep them viable.

In Tepoztlán, the process of absorbing the local economy into regional and national ones, in addition to the inflationary pressures of tourism, is gradually destroying the agricultural base of local inhabitants. Still, their future could be made more viable with the development of small local industries, the improvement of marketing infrastructure in rural villages, investment in crops for which there is local demand, and policies that make it possible for poor households to obtain credit, technical assistance, irrigation, and land. In Jaral del Progreso, a very different pattern of impoverishment has emerged through the process of agricultural modernization. In this município, poor households could expect a better future if they had more access to irrigation, land, and markets and if the gains from industrial employment could be invested in agriculture and small local industries, some of which are linked to agricultural activities and others of which are linked to the industrial base in nearby cities. The lack of potential for agricultural development in Unión de San Antonio has stimulated a strong migratory tradition that has been important not only in the economic growth of the município but also in the gradual destruction of alternatives to migration. Nevertheless, the future of Unión could be built on the basis of its already developed linkages to larger industries in nearby cities in Jalisco and Guanajuato and through the intensification and modernization of dairy ranching, a development that would require changes in government policies about land utilization and access to credit and technical assistance. In Villamar, the future may present almost intractable problems, not primarily because of the inherent poverty of the region, but because of the lengthy tradition of dependence that has developed. If such a tradition could be altered, local agricultural production might be stimulated through infrastructure that would make labor-intensive production of fruits and vegetables viable and the município could more effectively be linked to regional sources of employment in agriculture and industry.

Thus, in each município, a combination of policy changes and specific investments hold some promise for increasing the extent of regionally or locally available employment. An important question, then, is where the initiative would come from if local and regional economic opportunities were to be expanded in the four municípios. Some problems they share require policy changes, others require investment capital, and still others require strong leadership. The problems of Tepoztlán, Jaral del Progreso, Unión de San Antonio, and Villamar are complex and so are the possible solutions to the economic and political constraints they face. The next chapter considers possible policy, organizational, and local responses to these problems.

5

The Politics of Rural
Development in Mexico

For Tepoztlán, Jaral del Progreso, Unión de San Antonio, Villamar, and hundreds of rural communities like them, the future will be bleak unless significant efforts are made to encourage the creation of employment opportunities in rural and semirural areas. Such efforts will begin to respond to the needs of rural households that currently search diligently for the means to ensure their subsistence and increase their income. A strategy of rural development focused directly on the creation of productive sources of employment and income also promises to help Mexico create a better quality of life in its large and congested urban areas by providing alternatives to rural-to-urban migration. For rural and semirural areas to grow and prosper, a series of important changes must occur. As a first step, of course, the Mexican government must make employment a principal goal of its approach to rural development, changing its orientation from the traditional and almost exclusive focus on increased agricultural production.

Making employment creation a priority in Mexico is more easily suggested than accomplished, however. The rhetoric of government plans for rural regions has stressed production, employment, and social welfare goals for many years (see, for example, Mexico 1983, 1985; Oficina de Asesores del C. Presidente 1980). In spite of this rhetoric, any ongoing interest in rural areas within the government has been generated and maintained by concern at the national level about declining food production and rising import bills for basic food commodities (see Grindle 1982; Spalding 1984). In this sense, urban and industrial interests defined the goals of government policy toward the rural sector, a characteristic that remained true through the mid-1980s. Although poor rural inhabitants have consistently voiced interest in access to land and employment opportunities, rural development initiatives have just as consistently focused on engendering technological modernization in farming practices and stimulating the production of corn and beans, goals that accorded with national priorities in agricultural production and food self-sufficiency (see Grindle 1985a; Spalding 1984). In general, to the extent that job creation has been a formal policy objective, it has been sought through emergency public works projects that have had little permanent impact on longer-term productive employment and have required none of the politically difficult changes that are implied in the alteration of existing economic policies (Schumacher 1981). Thus the objectives sought by state planners and politicians often diverge from or even conflict with the concerns of rural inhabitants.

This chapter examines the politics of policymaking and implementation in an effort to determine how important changes in long-standing government practice can be brought about in Mexico. The urban and industrial bias of prior development strategies is probably a given for future plans and policies. Thus the issue for bringing about significant reform is how to mobilize urban concern for conditions in the countryside. In the case of Mexico, where the policymaking process is highly centralized in the executive bureaucracy, state planners and decisionmakers must become convinced that the future development of the country requires an employment-focused approach to rural development. They must believe that this development strategy is worthy of becoming a goal for reformers, a project for bringing about policy change. In the following pages I consider first the issues presented by policy and organizational reforms and then the characteristics of the policy process in Mexico to see how a reform "project" might be developed. Finally I analyze previous experiences with rural development strategies for the lessons they offer would-be reformers in the future.

What Must Be Done?

If Mexico's rural areas are to begin providing more income-generating opportunities, important changes must be made in terms of national development policies, control over public resources, and mobilization of local resources and initiatives. At the level of macroeconomic and sectoral policy, the Mexican state must assume a significant role in providing a more positive environment for rural development. In addition, state and local governments require greater control over decision-making and budgetary processes if they are to become more active in generating sources of employment for local communities. Local initiatives are also important for their ability to respond to local needs, resources, and opportunities at the same time that they remain accountable to the concerns of local inhabitants. These changes, considered in greater detail below, face serious obstacles within the context of Mexico's political economy.

A National Policy Environment for Rural Development

The Mexican state has a large role to play in stimulating rural development through its control over the policy environment and the context for institutional innovation (see, for example, Byerlee et al. 1983). Development specialists have become increasingly aware that aspects of national fiscal, monetary, and trade policy often have a greater impact on the ways in which rural areas develop or underdevelop than a plethora of specific programs and projects directly focused on rural communities (see, for example, IDB 1986:83–84; Page and Steel 1984; Timmer, Falcon, and Pearson 1983; World Bank 1986: 61–109). Thus, for example, they have argued that, "especially in the long run, macroeconomic policy determines the rate of growth of both urban and rural sectors and also conditions the structure of that growth. In particular, the degree of job creation and the distribution of income are more a function of macroeconomic policy than of sectoral investments and project design" (Timmer, Falcon, and Pearson 1983:218).

The impact of price, trade, exchange rate, and interest rate policies appears to be particularly important for rural areas. Policies adopted in Mexico and elsewhere to encourage industrialization have distorted the prices received for exported and domestically consumed agricultural products and have raised the costs of agricultural inputs. This generalized bias against agriculture has been exacerbated by an overvalued exchange rate and a high rate of inflation in recent years, lowering the cost of agricultural goods to consumers and implicitly taxing the sector. Rural-urban terms of trade have been significantly affected by the impact of the overvalued exchange rate, high domestic inflation, protection of domestic industries, and subsidies to consumers (see World Bank 1986: chap. 4). The impact of these policies was offset to some degree for large commercial farmers—who have long received a variety of generous subsidies from the government—but not for peasant producers. Sectoral policies biased against rural small and medium-scale activities have also inhibited agricultural and rural development (see Haggblade, Liedholm, and Mead 1986; Page and Steel 1984). Pricing policies that have reduced the income of farmers growing staple crops, for example, have also limited the incomes of rural households, encouraged crop substitution, and stimulated migration when other income-generating activities were not available locally (see Chapter 3).

Interest rates—significantly influenced by government policy in Mexico—have an important effect on the availability of credit in rural areas and its accessibility to various categories of borrowers (see Von Pischke, Adams, and Donald 1983). This effect has been important in shaping the rate of investment. Interest rate policies are also important in rural areas in terms of influencing voluntary savings for investment (see especially Adams 1983; Meyer 1985). In both cases, the Mexican experience has been adverse to the interests of small-scale and rural producers of agricultural and nonagricultural goods. In addition, tax codes can provide important incentives for altering future investment by medium and large-scale enterprises, although small-scale activities are generally unaffected by them. Again, tax policies in Mexico have affected the location and scale of industrial enterprises in the country and in the process have had an adverse effect on the rural economy.[1]

The economic crisis of the 1980s created an opportunity for introducing long overdue reforms in the structure of incentives and opportunities in Mexico by adjusting the exchange rate to a more realistic level, altering rural-urban terms of trade, and eliminating a number of costly subsidies on consumer goods. In particular, successive devaluations of the peso significantly altered the nature of incentives for imports and exports of goods and services. Moreover, in the aftermath of new

1. In the 1980s the Mexican state undertook initiatives to create a more encouraging environment for investment in productive activities. Trade policy, for example, was liberalized in 1985 to encourage industrial expansion for both exported and locally consumed goods (*BLA*, February 27, 1985:65–66; *LAM* 1986:63–66). Foreign investment policy was also altered in an effort to make the country more attractive to investment from the United States and elsewhere. Priority industries offering special incentives to foreign capital included agro-processing and capital goods industries. Priority zones for such investments continued to ignore the rural central plateau area but took the positive step of excluding urban agglomerations such as Mexico City, Guadalajara, and Monterrey (*LAM* 1986:59). Significant efforts to eliminate bureaucratic regulation of foreign investment were also introduced (*LAM* 1986:59–60). In addition, a major program introduced tax credits for firms that create jobs, produce necessary goods, purchase Mexican equipment, and locate in priority zones. In this program, special consideration was given to small and medium firms that make new investments. The certificates for fiscal promotion, or CEPROFIs, were an outgrowth of the López Portillo industrial development plan (see *BLA* 1979:122–124). Under the de la Madrid administration, CEPROFI tax credits were made available to small and priority firms willing to move out of the Mexico City metropolitan area (see *BLA*, February 13, 1985:49, 55).

austerity measures introduced in 1986, subsidies on consumer prices of basic commodities were reduced even in the face of significant political concern about the resulting economic impact on low-income urban consumers. Nevertheless, efforts at economic liberalization could not counteract the impact of continued inflation and a dramatic drop in international oil prices in late 1985 and 1986 that cut oil export earnings by 57 percent (Cornelius 1986:4–7). Inflation continued to reach 80–100 percent annually. Increasingly, observers were skeptical of the government's ability to sustain such reforms (Cornelius 1986:28–32). The administration of Miguel de la Madrid was strongly criticized for opening up the economy to increased foreign investment, for sacrificing concerns for social equity, and for failing to pursue measures that would bring relief from the growing external debt (Cornelius 1986:28–29). Others argued that the economic liberalization and destatization measures did not go far enough to create a positive environment for investment (Cornelius 1986:29). There continued to be a lively and often acrimonious debate in Mexico between proponents of a neo-liberal market-oriented development strategy and those who wished to see a continuation of the mixed economy model that had characterized the country's development in the past (Cornelius 1986:30–31). As long as the debate remained unresolved, so did the issue of the extent to which major macroeconomic reforms could be sustained.

Prospects for public and private sector investment in rural development were significantly affected by the economic crisis also. In 1982, Mexico was unable to service its foreign debt of more than $62 billion. Expansionary government spending in the 1970s, encouraged to tremendous proportions during the oil boom years of 1977–1981, was a significant factor in the debt crisis (see Wyman 1983:12). Mexico's problems continued to mount in the 1980s, and after lengthy negotiations with the International Monetary Fund, the government agreed to cut back drastically on public expenditures. A large number of programs were subsequently affected by austerity measures. Domestic and international conditions suggested a considerable period of economic difficulty before even modest recovery could be expected. In the meantime, public budgets for major new commitments to rural development were highly unlikely. In fact, given government austerity measures and political concern about the volatility and loss of political support in urban areas, decline in the availability of rural-oriented public goods and services, even including maintenance of existing programs and infrastructure, was a strong likelihood. The national development plan unveiled by the administration of Miguel de la Madrid highlighted the constraints on public investment when it called on Mexicans to achieve "more with less" (see SPP 1983b).

A similar situation characterized the environment for private investment in Mexico. In the immediate aftermath of the economic crisis of 1982, such investment slumped by 25 percent in 1983 and new direct foreign investment declined from 1.4 billion dollars in 1981 to 709 million dollars in 1982 and 374 million in 1983 (*LAM* 1986:60). New investments stagnated at this level in 1984 and 1985. Capital flight was massive; the World Bank estimates that, between 1979 and 1982, more than $55 billion left the country (World Bank 1985:64). In the 1980s, the Mexican state attempted to court private investment, taking advantage of its control over the policy environment to provide tax incentives and other inducements to attract capital for high-priority goals. Nevertheless, prospects for a rapid or even medium-term recovery in the Mexican economy remained poor, although the crisis did encourage greater foreign investment in labor-intensive productive activities because of changes in the relative costs of labor. Thus it appeared that in the mid-1980s the Mexican state had the potential to make rural development a national priority but had little capacity to invest large amounts of capital in rural areas to support this priority.

Decentralization: Power and Administrative Control

Inhabitants of Tepoztlán, Jaral, Unión, and Villamar share an important perspective with rural people all over Mexico: they feel—with good reason—that they have little control over or access to a plethora of decisions that affect their daily lives. Decisions about public and private investments, political representation, and access to a variety of economic opportunities are made in distant capitals and in negotiations from which they are excluded. As we saw in the research communities, the typical response to exclusion from political decisionmaking has been to search for economic solutions to household problems. One way to encourage rural inhabitants in Mexico to find a larger political voice is to bring the decisionmaking process closer to them, to the offices of the state and local government officials to which they have somewhat greater access. Decentralization of decisionmaking authority and responsibility is a step in this direction, even though it is an extremely difficult step to take.

State and local governments in Mexico are weak in terms of their control over financial and human resources. The Mexican government that emerged from the Revolution of 1910 is a highly centralized one.

To achieve and maintain this situation of political and economic control, state and local governments have been systematically starved of resources and have also been denied a significant role in policy and/or program initiative. Historically the result has been a system in which "each successive level of government is weaker, more dependent, and more impoverished than the level above" (Fagen and Tuohy 1972:20). In recent periods state and municipal governments have accounted for a declining proportion of public spending. In 1970, for example, state governments accounted for 12.1 percent of all public spending; by 1980, this proportion was 8.8 percent. Conditions for municipal governments have been especially penurious and growing worse; in 1970, municipalities accounted for only 1.6 percent of all spending, and by 1980 this figure had contracted to 1.1 percent of public spending (SPP 1982:455). Until the 1980s a series of government policies to stimulate decentralization could not alter the long-standing dependency relationship of the 31 states and nearly 2,400 municipalities on the federal government. A long history of small budgets and poor career mobility opportunities also left state and local governments starved for qualified and committed human resources in rural development and other fields.

Nevertheless, state and local governments are not completely excluded from the policy process. In fact, state and local authorities are regularly assigned important responsibilities for maintaining the political and social peace at local levels, for intervening in resource allocation decisions, for providing information on local power relationships, and for distributing jobs and contracts with an eye to building political support. Some evidence of at least nominal commitment to the idea of decentralization was made by the de la Madrid administration in the early 1980s. At that time, state development planning committees, bringing together officials from local, state, and national levels serving in the same state, were given the responsibility for plan and program initiatives in the state, in the expectation that greater regional equity would be promoted at the same time that administrative and fiscal decentralization was to be achieved (see SPP 1983b:175). The power to determine how and when state and local governments will participate, however, remains concentrated in national political hands (see for example, SPP 1983b: chap. 9). The unequal division of resources and power between the central government and subordinate levels has deep historical roots in the twentieth century, responding to intense political exigencies experienced by generations of national leaders. If state and local governments are to play a more active role in initiating and supporting more employment-oriented rural development, the necessary political "space" will have to come from national political leadership.

Stimulating Local Initiative

The most creative aspect of rural development is the potential to put local resources and abilities to work around productive opportunities. Needs exist, opportunities exist, and interest in making them work exists at the local level. The possibility of pursuing rural development objectives through local initiative has an immediate appeal for many development specialists and others concerned about participation. These individuals have long argued that locally based initiatives provide the flexibility, responsiveness, and opportunity for effective participation that is so often missing in top-down, state-led programs and projects (see Cohen and Uphoff 1977; Esman and Uphoff 1984; Gran 1983). Moreover, community-level initiatives ensure that there will be the local definition of needs and local responsibility for pursuing them. It is easy to move from these perspectives to the view that local initiatives can readily provide the important characteristics that are so problematic for state-led efforts.

In spite of such enthusiasm, a locally initiated approach to rural development in the Mexican context must begin with modest and politically realistic expectations. Local communities in Mexico are often highly factionalized, frequently around concrete economic grievances—the distribution of land and jobs, for example—that are played out in local politics and are linked to regional, state, and national politics through the strong brokerage role of the PRI (see, for example, Schryer 1980). A large number of rural communities are also becoming more highly stratified in terms of the economic resources controlled by different individuals and households (see Dinerman 1982; de la Peña 1981). This statement is particularly true of rural communities in the central plateau area where a cash economy and links to domestic and international markets have long had differentiated impacts on local inhabitants. Ejidos are often characterized by highly contentious internal divisions relating to issues of power (who's in and who's out in ejido management), land distribution, and unequal control over economic resources (DeWalt 1979; Ronfeldt 1973; Warman 1980). In the past, many local initiatives foundered on the question of factionalism and distrust. Indeed, fear of dishonesty, of being swindled, of being taken advantage of by their neighbors have empirical referents for a large number of rural inhabitants.

The implication of local factionalism for organizational efforts is simply that any initiatives are likely to be affected by local conflicts, many of which may have little to do with the specific goals of rural development. Because local communities are often factionalized, local initiatives will include some people and exclude others; moreover, it can be anticipated that there will be attempts to appropriate resources for factional aggrandizement, and jealousies and tensions will be exacerbated. In addition, because local communities are often stratified, local initiatives will undoubtedly benefit some more than others, possibly even deepening the extent of class differentiation. At times, equity concerns may be more easily implemented by authoritative agents from outside the community (Leonard 1982). These are not reasons for rejecting the possibility of local initiatives, but they indicate some reasons for caution in expecting local inhabitants to be able to identify common interests easily, to work together without friction in pursuing them, or to benefit similarly from joint efforts.

Moreover, enthusiastic and independent peasant response to state-initiated programs and projects cannot be assumed. Local organizations in Mexico have a long history of having been manipulated, coopted, or threatened by the political institutions of the state—the PRI, the National Confederation of Peasants (CNC), or its local and regional affiliates (Anderson and Cockcroft 1972; de la Peña 1981; DeWalt 1979; Ronfeldt 1973). Similarly, Mexican peasants have often been defrauded, swindled, abused, or manipulated by the state institutions and their agents who are responsible for rural development initiatives (see Grindle 1977a:147–163). In practically every community, there are vividly recalled examples of official credit agencies' overcharging local inhabitants for services, of state-supplied fertilizer's arriving too late for the harvest or seriously altered in content, of state marketing agency officials' underweighing the grain delivered to them by peasants, or of the PRI's determining who should be beneficiaries of state services. This legacy has created considerable local resistance to cooperation with state-initiated schemes. Local inhabitants, even when organized independently, will consider it risky business to collaborate too closely with the state agencies that offer the goods and resources important for supporting local initiatives. As one public official with experience in community organization stated, "Of course [the peasants] are distrustful; they have reason to be. Government and party officials have come over and over again making promises which they never keep and at times deceiving and taking advantage of the people" (quoted in Grindle 1977a:147).

Because of a history of disappointing local initiatives and considerable evidence of personalism, dishonesty, and exploitation, initiatives at this level in the future should begin modestly, speak directly to immediate economic interests, and offer tangible incentives to participants. That is, programs relying on local initiative should entail only modest amounts of risk to those who are expected to become active in them. This is in fact the consensus view of many development specialists who are concerned with issues of local organization and participation in a wide variety of country contexts (see, for example, Esman and Uphoff 1984). The advantages of flexibility, responsiveness, and participation that are regarded as positive aspects of locally initiated activities are real, but so too is the fragility of local organizational forms that can achieve these characteristics.

The Politics of Reformism in Mexico

Alterations in national development policy, decentralization of decisionmaking power, and the stimulation of local initiative are three reforms that would encourage rural development in Mexico. It is not an easy task in any political or economic system to make such important changes. It requires a group of reformers and decisionmakers who are convinced of the need to pursue a reoriented policy and who have taken on the task of change as a project. These officials need to have a good sense of the political and economic environment in which policy reform will be sought. Reformist initiatives need to be engineered through frequently resistant bureaucracies and past the opposition of the beneficiaries of prior policies. Managing opposition is only part of the task, however. Support for new policy directions must also be mobilized within the government and outside it. Not insignificantly, time for experimentation with new solutions and evaluation of their consequences needs to be available. Although these are requirements for achieving policy reform in any political system, the specific characteristics of Mexican policymaking both facilitate and hinder the successful introduction of change.

In several ways, reformist initiatives can be introduced more rapidly and coherently in Mexico than in many other political systems. Mexico has a decisionmaking process that is centered in the executive branch at the national level (see Grindle 1977a; Purcell 1975; Ward 1986). The legislature, despite increased representation of opposition parties in the 1980s, is largely uninvolved in policymaking except as a legitimizer of decisions made by the president and other high-level officials. The dominant party, the PRI, is not actively involved in policy formulation; its principal function in the system is to maintain political control and to mobilize support for executive decisions. Typically, public policy in Mexico is made by high-level presidential advisers who rely on a large cadre of middle-level technocrats for information and advice. Presidential advisers and technocrats are often well qualified in particular areas of expertise. In addition, they are frequently highly sensitive to the realities of bureaucratic politics in Mexico because they may have spent their careers moving about the complicated labyrinth of the public sector (see Grindle 1977a; Smith 1979). Once a consensus or coalition of agreement about policy has been reached within the executive branch, these same officials have become accustomed to relying on the considerable power of the presidency and the loyalty of the PRI for its adoption (Levy and Székely 1983: chaps. 3 and 4; Purcell and Purcell 1980; Ward 1986:40–41).

Moreover, bureaucratic officials have considerable scope for designing coherent responses to the problems they perceive. The closed and executive-centered process of decisionmaking affects the extent to which organized groups and interests have access to the planners and decisionmakers. Thus in the formulation of any particular policy, the direct participation of politically important groups may or may not be present; decisions as to who is to be consulted and whose opinions are to have influence are generally made by public officials, not by external groups that have demands to make and pressures to bring to bear on the government. The first public communication about policy reform often comes when it is announced officially—usually by the president. I do not mean that public officials in Mexico make policy in isolation from societal pressures. Instead, there is often extensive private and informal consultation with groups whose interests are affected by policy reform. Nevertheless, generally the planners and decisionmakers control access to the policy process and exercise discretion as to who is to be consulted or informed. Even politically important interests are cast more often in the role of reactors to government policy than as actors in its formulation (see Purcell 1975). As I will suggest later, this characteristic of the policy process means that implementing policies in Mexico tends to be much more difficult than formulating them.

The policies that emerge from the government are highly responsive to presidential perspectives, and policy change is facilitated by the regularity of administration change in Mexico (see Grindle 1977b). Every six years, a new administration takes over extensive executive power. All presidents are members of the dominant party, but each takes office with his own team of cabinet officials and high-level advisers; these individuals in turn select their own cadre of subordinates, who in turn select their personal team and so on down through middle ranks in the bureaucracy. Ultimately, this system of succession and appointment centralizes great power in the hands of the president, power that deserts him as soon as he relinquishes his position to a successor. In the interim, presidents are able to make substantial changes in the direction of policy because few within government have an independent base of support that would allow them to take a position in clear opposition to his perspectives. Presidential statements are assiduously studied for evidence of his policy preferences. Presidential appointees interpret their own career chances in terms of how well they are able to respond to his concerns or to link their own policy preferences to his broad goals and perspectives (see Grindle 1977a). Policy changes tend to be concentrated at the outset of an administration, when the incentives to follow presidential leadership and to differentiate new initiatives from prior policies are strongest. Thus after an initial period of planning and discussion, a plethora of changes is usually announced early in the second year of a presidential administration. In addition, in the 1970s and 1980s, a trend toward increased presidential autonomy in the last year of the administration emerged as a characteristic of Mexican policymaking.

Despite such extensive presidential and executive power, however, there are also obstacles to policy reform in Mexico. Within the government, there can be considerable disagreement about the appropriate "line" of policy in cases in which the president does not assume a role of leadership. In such cases, consensus or coalition building within the bureaucracy is a prerequisite to policy change. This kind of coalition or consensus is not easily put together, because institutional rivalries are strong and divisions are exacerbated by the interest of each in attracting presidential interest at the cost of others. At times, however, agreement can be forged on the basis of complementary diagnoses of the problem or on the basis of a shared constituency of agency or ministry clients (see Grindle 1977a). The mobilization of support within the bureaucracy must be accompanied by the marshaling of strong and sustained presidential support for a new initiative; without it, few reforms have the capacity to survive the political pressures that are brought to bear on the implementation process.

Moreover, the quiescence of political and economic groups outside the government is essential to the maintenance of the current regime.

Organized groups of industrialists, commercial and export interests, certain organized labor unions, and a network of caciques are among those whose interests are usually taken very seriously—even if their opinions are not overtly solicited—in any new policy initiative. Policymakers who wish to see their reforms survive are therefore well advised to keep the changes they propose within the bounds of acceptability to these interests or at least to be certain that proposed changes will not lead to coalitions of opposition among them. In particular, the economic power of entrepreneurial groups in industry, agriculture, and finance is great enough to give them considerable capacity to obstruct initiatives they oppose, if not when policy is made, then at least when it is pursued. Thus under normal conditions, policymaking in Mexico proceeds with a tacit recognition that certain types of changes, such as an expansion of the agrarian reform, cannot be considered.[2] This political sensitivity to established interests is least evident at the end of a presidential administration when the "space" for major changes tends to expand (see Smith 1984b).[3] During most of an administration, however, and especially when particular interests have been beneficiaries of prior development policies, bringing about changes that will affect them negatively is difficult at best.

Public officials favoring reforms in Mexico must also be concerned about mobilizing support beyond the bounds of the executive branch. To some extent, once presidential support has been secured, they can count on the approval of the PRI and can attempt to utilize the official party to obtain broader public acceptance of reform. This simplifies their task, but time has taken its toll on the capacity of the PRI to deliver effective support. Frequently, announcements of new initiatives in public policy are met by passivity on the part of a large number of citizens. Such a response reflects the practice of political participation in the country. The political system of Mexico has, since the 1940s, sought consciously to demobilize and disaggregate any independent political initiatives at the same time that it has sought to incorporate organized groups into the official party in order to control their activities (see, for example, Hellman 1983; Stevens 1974).[4] In short, the government

2. Such "off limits" policy options, present in all political systems, are discussed in Bachrach and Baratz (1970) as "nondecisions" that perpetuate systematic biases for different social groupings.
3. The expanded political space for Mexican presidents at the end of their administrations was evident under the presidencies of Luís Echeverría and José López Portillo. As an example, López Portillo nationalized the banking system just four months before he left office.
4. The administration of Miguel de la Madrid engineered a system of popular consultations, part of a National Democratic Planning System, in which various policies and programs

almost always has a captive and verbal support group for new policy initiatives; on the other hand, such support rarely runs deep, is often pro forma, and is accompanied by considerable political cynicism. Such attitudes can prove of little assistance if the government is concerned about pursuing policies that require the active involvement and commitment of large numbers of people. In such cases, the obstacles to policy implementation are considerable.

Indeed, implementation can be a lengthy process of political participation, accommodation, and negotiation in Mexico (see especially Grindle 1980). People who have been excluded wholly or partially from the policy formulation process often find it much easier to bring pressure to bear on the public officials who are charged with day-to-day responsibility for allocating policy, program, and project resources. As a result, considerable "slippage" often occurs in the ways a policy, program, or project is pursued after it has been announced. This disparity tends to widen the gap between the rhetoric of policy objectives and what is accomplished in fact. The more complex the administrative requirements for pursuing and monitoring reform, the more likely that "slippage" will occur. Thus the frequent failure to achieve stated goals can often be traced to extensive accommodation and reallocation of resources during implementation. Public officials tend to be aware of this characteristic of the policy process; they are also aware that the maintenance of the political peace may require that accommodation of important interests and adjustment of objectives for specific cases be accepted as a routine part of the overall policy process (see Grindle 1980). In fact, the impressive stability of the Mexican political system can be linked, in no small way, to the politics of the implementation process that minimizes the number of "losers" when a policy change is adopted. Clearly, however, this condition places constraints on the ability to pursue far-reaching reforms.

The particular requirements of rural development policy bring about additional difficulties within the Mexican political system. It will be recalled that the policy changes important to stimulating employment require not only macroeconomic changes but also considerable decentralization and local autonomy for solving problems of rural underdevelopment at the regional and local level. Political and administrative centralization have been achieved at considerable cost in Mexico; they

---

were to be discussed with affected groups (see Ward 1986:44, 52–53). In fact, the *consultas populares* have been used primarily as public forums for generating support for policies already designed and for informing the public about them.

are not lightly relinquished. Since the early 1970s, emphasis has been accorded to administrative decentralization in government plans, and much rhetoric has been devoted to such efforts (see, for example, SPP 1983b). As of the mid-1980s, however, little had actually been achieved and, if anything, states and local governments were poorer and less powerful in 1987 than in 1970.

Local autonomy is an even more difficult change in Mexico. To a large extent, the maintenance of political stability in Mexico has been achieved through a well-controlled system of patronage, cooptation, and repression that reaches down to the most local level. The Mexican political system has long been noted for the sophisticated way in which public resources—jobs, a piece of land, access to credit, preferential treatment in bureaucratic encounters—are used for the purposes of coopting potential or actual sources of dissent or opposition or simply to pay off loyal government supporters (Anderson and Cockcroft 1972; Cornelius 1975; Hansen 1971; Stevens 1974). The system has been particularly effective in rural areas and is an important reason why peasants, who have benefited least of all economic sectors from Mexico's historically impressive economic development, have remained the most enduring support groups for the regime (see, for example, Reyna 1974). Support is maintained through a series of local bosses, or caciques, who trade allegiance to the political system in return for the economic and political benefits that ensure their continued ability to control local clienteles. This is one important way in which the implementation process is used to maintain the political peace in the countryside. Public agencies converge on the peasant in their control functions. Indeed, public officials charged with day-to-day allocation of resources are well aware that part of their job is to make certain that local powerbrokers are well rewarded and that local economic elites are not unduly threatened by development projects (see Grindle 1980). In this effort, they are helped by the PRI, whose local level network is essential to keeping the political peace in Mexico's rural areas.

Granting more local autonomy would mean the loosening of these bonds of political control and the possible emergence of local unrest or independent political activities. One planner, for example, commented on a state governor's reluctance to cooperate with a particular rural development initiative because of its potential to be politically disruptive. "He is afraid, when we begin talking about putting decision making power in the hands of the peasants, that this is going to create all sorts of expectations that he is not going to be able to meet. He thinks first in terms of political stability." The government would not be likely to remain committed to locally managed solutions to development problems for long if political repercussions became noticeable.

A particularly thorny problem for reformists concerned about employment and rural development is the lack of a mobilized constituency pressing for policy change. Although in general the initiative for policy reform comes from the executive in Mexico, this pattern is particularly strong in terms of how rural development problems affecting the peasant population are addressed. As I have argued, peasants are not organized to make effective demands on the political system. Moreover, as was evident in the four município profiles in Chapter 4, rural inhabitants in Mexico have responded to worsening economic conditions by diversifying their economic activities and having more frequent recourse to migration, not by organizing politically. There is little reason to expect that public officials in the future will find themselves any more pressured from below than they have been until the present to address the problems of rural employment.

In summary, the centralization of political power in Mexico can facilitate the task of the reformer particularly if the president is a strong proponent of the policy change. At the same time, however, once a reform has been adopted, it often faces serious obstacles that will impede its effective implementation. Particularly in the case of new rural development initiatives, it can be difficult to mobilize sufficient public support to help overcome the resistance of commercial agricultural interests, rural caciques, and regional governments. Given the characteristics of political participation in Mexico, rural pressure for change is not likely to be effectively channeled to the national elites who make decisions about the country's development strategy. If change is to occur in Mexico's orientation toward the countryside, it will have to find significant support among these elites. It is therefore important to ask how their interest and concern in rural employment problems can effectively be mobilized.

Elite concern for problems of the countryside has become notable in the past, however. The concern for rural development that became evident in the 1970s, for example, resulted from a consensus among public officials in certain ministries and agencies that peasants were important to national development strategies because they were the primary producers of critical staple crops. This concern led to a series of initiatives focused on increasing production among smallholders and ejidatarios—in effect, the rediscovery of the importance of peasant production led to the emergence of a rural development project among reformers, even in the absence of a strongly mobilized peasantry pushing for reform. Similarly, greater concern about employment in rural development must emerge from within the policymaking apparatus if reform is to occur.

Elites who make the important choices about future development strategies do not operate in a historical vacuum, however. The choices they make are influenced by their understanding of the experience of prior policies and their perceptions of what the appropriate role of the state should be. These are important factors influencing elite perceptions in Mexico because of the leadership the state has assumed in rural development strategies in recent years and the political and economic realities facing a historically activist state in the 1980s. The creation of a new project focused on rural employment will therefore be affected by the legacy of past rural development initiatives and by the changes in the role of the Mexican state caused by the economic crisis of the 1980s.

Lessons from the Past for Reformers

The Mexican state has traditionally been an activist one. Throughout most of the twentieth century, it assumed leadership for stimulating and guiding national economic development and for delivering on promises to the masses of the population in the areas of social, economic, and political rights (see Bennett and Sharpe 1980; Hansen 1971; Hellman, 1978; Purcell 1981). In the course of its history, it built up an impressive array of ministries, agencies, and parastatal organizations that are active in the formulation and implementation of economic and social policies and in the management of enormous numbers of programs and projects that affect the daily lives of millions of Mexicans. Similarly, the state acquired an extensive body of legislation that gives it the legal authority to intervene widely in the economy and to shape important social relationships in the country (Bennett and Sharpe 1980; Purcell 1975). Equally important to the power of the Mexican state is the PRI, which served to integrate a wide range of interests into a dominant party structure and to control participation, dissent, and opposition within the political system (see especially Hansen 1971). Combined, these factors enhanced the statist orientation of development in Mexico and strengthened the government's activist role.

In the case of rural development, the historical role of the state has been extensive and formative, as has been suggested in previous chapters. Through the massive implementation of an agrarian reform in the 1930s and the development of large irrigation districts, it established the ejido and the large commercial export sector as enduring characteristics of the country's rural areas (see Grindle 1986: chap. 4). The state determined the priorities for agricultural development through its investments in infrastructure, research, credit, and subsidy programs for large-scale commercial farmers (see Carlos 1981; Hewitt de Alcántara 1976). It also developed a large cadre of public officials with expertise in agricultural sciences, agricultural economics, and agrarian issues.[5]

Beginning in the early 1970s, the Mexican government introduced a number of initiatives directed toward ameliorating what it considered to be the major problems of rural development in the country. Each of these initiatives holds important lessons for reformers. Prior experience suggests that rural development initiatives: (1) are very expensive in terms of financial and administrative resources; (2) must address the problems of access to land and the question of landlessness in Mexico; (3) are significantly constrained by the quality of land and water resources available to producers; and (4) must be supported by changes at the level of macroeconomic policy. These lessons must be applied to new rural development initiatives. At the same time, reformers must face the particularly unsettling reality that, for the foreseeable future, few public resources will be available for dedication to rural development initiatives. The lessons of the past must therefore be applied to an environment very distinct from that which surrounded rural development efforts initiated in the 1970s and early 1980s. Central to this new environment is the lessening of the activist orientation of the Mexican state.

With the administration of Luís Echeverría (1970–1976), the Mexican government began to become more concerned about responding to what it perceived to be the problems of the country's rural areas, especially those areas that lacked irrigation and sufficient rainfall for highly productive agriculture. Three public sector efforts show how problems were addressed and what lessons were learned: the PIDER program (Program for Integrated Rural Development) was initiated in 1973; a rain-fed districts rural development initiative was introduced in

---

5. Recent figures provide some insight into the magnitude of the state's agricultural and rural development activities in the early 1980s. The Ministry of Agriculture and Water Resources and the Ministry of Agrarian Affairs managed a combined authorized budget of 147,693 million pesos in 1982, 14.7 percent of the federal budget. The Banco Nacional de Crédito Rural distributed an estimated 80.8 million pesos in 1983, whereas the state fertilizer company sold 1,872,750 tons of fertilizer (de la Madrid 1983:455–456, 449). The state-owned improved seed company produced 212,714 tons of certified seed for principal products, and the state marketing agency, CONASUPO, bought nearly 4 million tons of corn and beans from producers in 1982 and an additional 2 million tons in 1983 (de la Madrid 1983:447–448; USDA, personal communication).

1977; and a massively funded national food strategy known as the SAM (Sistema Alimentario Mexicano) was pursued between 1980 and 1982. Each of these programs was initiated through presidential leadership and was actively supported by a cadre of public officials who owed their positions and allegiance to the incumbent president. Each resulted from an urban initiative and responded to urban definitions of rural problems. Nevertheless, each program indicated a commitment on the part of the government to increase productivity in agriculture and to improve conditions of welfare in disadvantaged rural areas. Each also indicates the limitations on agriculture-based rural development initiatives and emphasizes the importance of considering employment creation to be central to the resolution of Mexico's rural dilemmas.

Government concern over rural development issues increased in the 1970s largely because of declining productivity in staple crop production, particularly that of corn and beans. Since the 1940s, Mexico had pursued a policy of import-substituting industrialization that relied on the agricultural sector to provide raw materials and generate foreign exchange to stimulate the development of urban and industrial sectors. A variety of policy instruments encouraged the productivity of export-oriented agriculture, particularly in the northern and northwestern sectors of the country.[6] The impact of these stimulants was significant, and export agriculture grew impressively in the 1960s and 1970s. The case of Jaral del Progreso reported in Chapter 4 is an example of the kinds of changes that modernization brought to Mexican production. Although the development of modern capitalist agriculture was encouraged by government policy, production for domestic consumption was largely ignored or discriminated against. In 1965, Mexico began importing, for the first time in many years, substantial amounts of its principal staple crop, corn. In the following years, balance-of-trade figures reflected an increased dependence on food imports. By the early 1970s, government planners had become increasingly concerned about the lack of production of staple goods and more concerned about dependency on food imports, particularly from the United States.[7]

Under the Echeverría administration, rural underdevelopment was interpreted in official circles as a consequence of the structural constraints on peasant agriculture, the sector of the rural population that

6. This was largely done through the creation of massive irrigation facilities, other infrastructure development, and subsidized credit, extension, research, energy, and mechanization.

7. The Echeverría administration was also concerned about rural poverty and its implications for political stability (see Grindle 1977a: chap. 4).

produced the major portion of the staple crops (see Grindle 1981). In particular, the official analysis focused on the role of intermediaries in the rural economy, those who extracted any available surplus from peasant production. To alter this situation, the Echeverría administration strongly affirmed the responsibility of the government to provide goods and services to peasant producers to free them from the exploitation of middlemen in the rural economy. The government proposed and then implemented a variety of measures designed to alter structural conditions in rural areas—higher support prices for basic staples, subsidized credit, direct marketing by government agencies, expansion of government storage facilities, provision of subsidized green revolution technology, improved transportation and communication networks, and expanded educational and health facilities.

A large number of programs were started to carry out these initiatives. One of the most visible was PIDER (see Cernea 1979, 1983; Grindle 1982).[8] The program was based on the development of microregions identified on the basis of indicators of rural poverty and potential for development. Each microregion—there were 139 of them by late 1983—served as a unit for planning an integrated set of projects for agriculture, physical infrastructure, health, sanitation, and education. Fourteen federal agencies were involved in PIDER's implementation; in 1981, primary responsibility for implementation was decentralized to the state level. PIDER enjoyed a heyday of official support under Echeverría, and in part for this reason it was relegated to lesser priority by the succeeding administration of José López Portillo, which was eager to establish its own policy identity; nevertheless, the program continued to absorb funds from both international agencies and the Mexican government. By 1983, it had accounted for $2 billion in funding (see Cernea 1983:1).

The López Portillo administration largely rejected the previous emphasis on structural constraints on the peasant economy and adopted instead an analysis that focused on technological constraints (see Grindle 1981). López Portillo and his advisers were visibly alarmed by increased dependency on food imports and the rising foreign exchange bill implied by this dependency, and they saw in technological innovation the means to improve productivity among peasant farmers without creating political tensions through attacks on the intermediaries in the rural economy. Planners in this period stressed that peasants were not

8. PIDER was an internationally visible integrated rural development project that received extensive support from the World Bank and the Inter-American Development Bank.

productive because the regions they lived in lacked basic productive and social infrastructure, and risk factors in subsistence and semisubsistence farming severely limited the rationality of adopting new technologies. Although the López Portillo administration's analysis of the problem was different from that of the previous administration, the two perspectives coincided in assigning the major responsibility for introducing rural change to the national government. As a result of the analysis, the government created the rain-fed districts program within the Ministry of Agriculture and Water Resources. The new districts, modeled on the idea of the irrigation districts established in the 1950s and 1960s to bring infrastructure and green revolution technology to those areas, were to serve as planning and administrative units for the diagnosis of production problems and the introduction of appropriate technological innovations in areas that did not have adequate irrigation.

The oil boom and vastly increased government resources made it possible for the López Portillo administration to introduce a much more ambitious approach to rural underdevelopment in 1980. This was the SAM, the Mexican food system (see Austin and Esteva 1985; Grindle 1985; Luiselli 1982; Meissner 1981; Redclift 1981; Spalding 1984). The primary motive for introducing the SAM was to achieve food self-sufficiency in basic staples and to increase access to nutritious food among low-income inhabitants by improving the efficiency of food production, processing, distribution, and consumption throughout the country. The SAM was an effort to deal comprehensively with the entire food system in the country and as part of this it sought, through the use of price and subsidy policies, to increase the attractiveness of staple crop production to farmers. Much of the emphasis in terms of production continued to center on technological innovation and the risk inherent in peasant productive enterprises, but the strategy, encompassing a variety of specific programs, gave considerably more attention to pricing policy than did earlier efforts. Each of its programs was expected to stimulate the production of corn, beans, rice, wheat, and oil crops by smallholders and ejidatarios, but subsidies were available for credit, inputs, crop insurance, and crop purchases to anyone who agreed to grow the priority crops. Between 1980 and 1982, the SAM strategy was estimated to have cost some 3 billion dollars in subsidies alone, and the government reported that the strategy accounted for more than 10 percent of the federal budget in 1981 (see Grindle 1985; Spalding 1985:1258). For a brief period, it contributed significantly to increased production of staple crops.

The SAM came to an abrupt end at the close of the López Portillo administration and the severe economic crisis of 1982. The impetus to accord high priority to the rural sector died almost as rapidly, in large part because of stringent austerity measures and preoccupation among government planners and decisionmakers with macroeconomic adjustments. Before this curtailing of official support for rural development initiatives in the 1980s, however, between the early 1970s and the early 1980s the Mexican government dedicated large amounts of money to rural development investments. State policies and programs expanded infrastructure noticeably, created networks of schools and health facilities throughout rural areas, provided more credit and extension services, and stimulated the adoption of green revolution technologies for agricultural production. At least until the severe financial crisis of 1982, rural development was a highly visible project for the Mexican government and one that received the attention of a significant number of public officials. PIDER, the rain-fed districts program, and the SAM indicate the extent to which coherent approaches to public problems can be generated through the policymaking process in Mexico. Because they were formulated within the executive branch by officials with special interest in their success, they were based on thorough analyses of the problems of Mexico's agricultural production. Public investments were then planned on the basis of persuasive technical analyses of the constraints on agricultural development in the country. Each reform was based on a critique of prior government development policies that affected the agricultural sector.

All three initiatives were concerned with the issue of agricultural production, but they were significantly different responses to this issue. The change in focus from one strategy to the next emphasizes the extent to which policy change is possible in Mexico. In particular, the initiation of a new administration or, in the case of the SAM, the rapid expansion of public resources, can lead to relatively rapid policy change. The negative side of this characteristic is that initiatives begun in one administration are altered considerably or even terminated with the change of administrations. PIDER continued to exist into the mid-1980s—but in much altered and diminished form—the rain-fed districts were largely forgotten, and the SAM was declared officially dead in early 1983.

PIDER, the rain-fed districts program, and the SAM also underscore the extent of centralization of policy planning and decisionmaking that exists in Mexico. They were considered to be national programs that would be funded and administered by the national government, with little direct involvement of the states or local governments. PIDER was eventually decentralized to the state level but only after its status had been considerably eroded through the process of administration change. A corollary of extensive centralization of decisionmaking in the programs was that the participation of rural inhabitants in their design and implementation was extremely limited. In the case of the PIDER, the peasant sector of the PRI was made aware of the program while it was being designed, although it was not seriously part of the policymaking process. In the cases of the rain-fed districts and the SAM, the programs were designed within a national government ministry or a high-level presidential advisory group, and when the plans were completed, they were announced publicly, with minimal advance notification to those who were expected to benefit from them.

Moreover, despite the level of resources dedicated to rural development initiatives in the 1970s and 1980s, productivity and conditions of welfare did not make lasting gains in Mexico during this period. In many ways, conditions actually worsened for the country's rural poor. As indicated in previous chapters, landlessness increased, the rate of temporary labor migration grew, the production of basic staples stagnated, and levels of insecurity became more grave (see also Grindle 1986: chap. 6).

Why did increased investment in rural development not bring more positive results in terms of productivity and welfare? The answers to this question are numerous. The programs required not only extensive financial resources but also extensive administrative commitment and coordination among official agencies. Their requirements in terms of the performance of a large and poorly coordinated bureaucracy were probably too great to lead to efficient implementation in the short term (see Cernea 1983). Institutional factors such as poor planning, lack of coordination among government agencies, inadequately trained personnel, competition over priorities among official agencies, the waxing and waning of the political popularity of specific programs, official corruption, and a multitude of similar problems clearly limited the impact of these rural development initiatives in Mexico (see Grindle 1982).

In addition, however, the discussion of Mexico's rural problems in the foregoing chapters provides some basis for questioning the approaches to rural development contained within PIDER, the rain-fed districts program, and the SAM. In particular, each program focused centrally on agricultural production as a preeminent goal. In this regard, the programs ignored the increasing reality in rural Mexico that a growing portion of the population is landless and has no access to productive employment opportunities on the land. Thus the programs had minimal, if any, impact on what has emerged as the central rural development issue: the provision of productive sources of employment for the large number of rural households whose livelihood no longer derives primarily or even secondarily from agriculture. The focus on increasing agricultural production also encouraged planners to over-look the important limitations on agriculture in some parts of Mexico caused by the poor quality of land and water resources. As we saw in the case of Unión de San Antonio, the natural resource base in some areas can be so poor that the economic feasibility of modernizing agricultural activities in them is questionable. As I suggested in Chapter 3, in some regions in Mexico it is essential to find alternatives to agriculture rather than to continue to increase investment in it. The initiatives of the past did not focus on this reality. In fact, although each program was based on a useful analysis of the constraints on peasant agriculture, none adequately responded to the fact that the terms "rural" and "agricul-tural" are not synonymous in Mexico.

Moreover, with the exception of some aspects of the SAM, the rural development initiatives focused at the program level on the delivery of goods and services to beneficiary populations. They tended to ignore macroeconomic issues related to interest rate, exchange rate, and tax and trade issues or policies that affect rural-urban terms of trade that would have altered the policy environment for rural development. There was some concern for pricing policy, especially within the SAM, but the reforms introduced relied principally on subsidies to both pro-ducers and consumers that could not be sustained without the capital generated through the oil boom or extensive international borrowing.[9] This was a principal reason for the rapid demise of the SAM when Mexico's severe economic crisis became apparent in 1982.

The rural development programs of the 1970s and early 1980s are notable for the extent to which the state assumed the role of leadership in assessing the problems of the rural sector, designing programs to respond to them, and directing the programs once they were put in motion. The statist orientation of the public sector was significantly lessened under the de la Madrid administration; moreover, austerity measures meant that the state had neither the economic capacity nor the administrative capacity to continue its strongly interventionist role.

---

9. Thus one observer reports that, "by the summer of 1982, CONASUPO was paying 8,850 pesos per ton for maize and selling this to tortilla makers for 1,000 pesos in a massive subsidy program that kept final tortilla prices less than half their real cost" (Spalding 1985: 1285).

The shift in emphasis was less evident in rhetorical pronouncements than in terms of activities actually pursued. Early in the administration of Miguel de la Madrid, the National Development Plan for 1983–1988 was produced. In it, the government reaffirmed commitment to the importance of rural areas in feeding the population, and it pledged to pursue the medium-term goals of social welfare, integrated agrarian reform, agricultural production, and the improvement of employment and income levels in rural areas (see SPP 1983b). Shortly thereafter, the National Food Program (PRONAL) was announced with significant fanfare; the government claimed that its strategy, programs, and projects would resolve long-standing conditions of underdevelopment in rural areas (see Mexico 1983).[10] Two years later, the National Integrated Rural Development Program (PRONADRI) made its appearance and echoed the government's commitment to the priority of rural development, especially in rain-fed areas (Austin and Esteva 1985; see Mexico 1985:16).

In fact, neither strategy was put into effect. The documents themselves survived as testimony to divergent concerns within the bureaucracy—the PRONAL was more concerned with peasant agriculture and the production of basic crops, and the PRONADRI emphasized the potential of commercial agriculture and livestock production. One clear reason for the divergent views expressed in PRONAL and PRONADRI and the fact that both proposals were left to languish was that the president was not effectively committed to addressing these sectoral problems. Given this situation, none of the individuals or ministries who put the programs together was willing to take the risk of pushing for a project that would bring little reward to the individuals or to their ministries or agencies.

Some have interpreted the disregard for rural development during the de la Madrid administration as a reflection of a common problem for sectoral policies during a period of macroeconomic adjustment. Thus one official responsible for formulating rural development policy stated, "This government can't really deal effectively at the level of sectoral policies because of the overriding preoccupation with the debt, the balance of payments, government deficits, and austerity."[11] Moreover, policymakers in the de la Madrid administration were more cautious about the role of the state in rural development. A planner argued, for instance, that "solutions to rural development problems are to be found

---

10. PRONAL was known as "son of SAM" because its motivating principles, although less ambitious, were the same as those of the SAM.

11. Interview, Mexico City, June 5, 1985.

in a combination of macropolicies and very localized packages of interventions that are specific and that are defined by the campesinos."[12] Another high-level decisionmaker echoed the lack of confidence in prior state-initiated efforts. "I have become convinced that the rural development projects like PIDER are extremely inefficient. They use large amounts of money and provide very few returns in increased productivity."[13] Finally, some argued that the state under de la Madrid lacked a vision of Mexico's future or effective control over events in the country that would allow it to pursue useful responses to rural underdevelopment. "The plans of this administration are all good ideas, but that's all they are, ideas. They don't have mechanisms to carry them through, they don't have vision, they don't have any substance. This is a coping state."[14]

Whatever the reason, it was clear that programs such as PIDER, the rain-fed districts program, and the SAM would not be pursued actively in the mid-1980s. The resources required for such strategies were not available, and the state was not in a position to manage such large initiatives. Moreover, influential members of the private sector and the strong influence of the International Monetary Fund on the adoption of public policies created an environment that was not supportive of state activism in the provision of goods and services to poor rural areas. Nevertheless, although the change in the role of the state placed constraints on the activities of would-be reformers, the difficult conditions of the 1980s may have created a propitious moment for assessing a national strategy for rural development. Concern with the macroeconomic environment and a more permissive attitude toward local initiatives could be one result of the hiatus in official involvement in rural development activities in the mid-1980s. These factors, combined with increased anxiety over the economic, social, and political problems of hyperurbanization could be significant factors in focusing official attention on the central role of employment in the future of rural Mexico.

Conclusion: Urban Mexico and Rural Development

Ironically, then, the urban bias and urban concern notable in prior rural development initiatives in Mexico may well continue to be the source of initiatives for rural development in the future. Urban Mexico

12. Interview, Mexico City, June 5, 1985.
13. Interview, Mexico City, June 6, 1985.
14. Interview, Mexico City, June 8, 1985. See also *New York Times*, June 25, 1985:1.

is not only the source of policy initiatives but also the focus of concern among political elites. Nowhere is this concern more evident than in Mexico City and the cities of the north of the country. Mexico City has an estimated metropolitan population of 16 million to 18 million people. It is probably impossible, both physically and economically, to provide adequate basic services to this agglomeration in terms of sanitation, water, electricity, schools, health services, roads, and urban transport for the foreseeable future. It is even more difficult to provide employment for such large numbers of people. Public concern about these issues is significantly influenced by the fact that conditions of life in Mexico City are increasingly difficult for the elites who make decisions about important policy initiatives: traffic congestion makes going to their offices a daily ordeal, often consuming two hours or more a day; the impact of pollution cannot be escaped, and many Mexico City inhabitants—elite and poor alike—have persistent "smoker's cough." Street crime affects policymaking elites directly, as it does their families and exclusive neighborhoods. Politically, those who make important decisions about development strategy in the country are also increasingly unsettled about the evidence of the decline in the legitimacy of the PRI and the emergence of more effective opposition parties.[15] This is the principal reason for elite concern about the rapid growth of the urban centers in the north of the country, where political subservience to Mexico City has always been fragile. Elections of the 1980s demonstrated the extent to which political opposition to the PRI could be mobilized by alternative parties (see *LAM* 1986:303–305).

These are urban problems. Nevertheless, it is increasingly clear that they have rural roots. The growth of Mexico City and the northern cities is a direct result of poverty and underdevelopment in rural areas. Moreover, as in the case of Mexico City, there may be no viable urban solutions to the massive problems caused by overpopulation and rapid growth. If this perspective is meaningful, then rural solutions can be viewed as more viable—solutions that make it possible for poor households to achieve decent standards of living in their regions. Increasing attention to rural areas would therefore be stimulated more by the impact of permanent rural-to-urban migration than by attention to the issue of temporary labor migration, the indicator of rural underdevelopment that has been stressed here. Nevertheless, to the extent that rural solutions come to be accepted as necessary to urban problems, then the

15. See, in particular, the discussion of 1986 elections in Chihuahua in *LAM* (1986:303–305). In these and other elections, the Partido de Acción Nacional (PAN) scored impressive numbers of votes, and there were widespread allegations of fraud against the PRI.

issue of employment in rural development can achieve much higher priority within policymaking circles. It should be emphasized that an appropriate rural development strategy will focus not only on the economic viability of rural villages but also on their links to the economies of regional towns and small cities. A central concern for employment will help maintain an initiative to make rural communities less dependent entities in the agricultural and industrial development of their regions.

The issue of international migration, which has been a focus of concern of United States foreign policy toward Mexico, is not a high priority among the elites who make decisions in Mexico (see Alba 1978; Craig 1981). Permanent and temporary migration to urban areas in Mexico is of much greater and immediate concern to them, for it continues a process begun in the 1940s and 1950s that has created the economic, social, and political problems of hyperurbanization. From this concern and the relationship between rural underdevelopment and urban migration might come the impetus for a new assessment of the role of employment in rural areas and in national development.

6

Rural Development, National Development, and the International Arena

Mexico's economic development in the decades after 1940 was oriented toward achieving growth through a strategy of import-substituting industrialization. Impressive growth in the economy was in fact achieved, with GDP increasing by an average of more than 6 percent annually (see Table 13). Although this strategy for economic development began to demonstrate serious limitations by the early 1970s, international borrowing and the discovery of massive reserves of petroleum allowed the country to continue to record high rates of growth in GDP without fundamentally altering the overall policy framework for development. Nevertheless, by the early 1980s, it was apparent to most observers that Mexican development policy had left a legacy of inefficient industries, a low level of effective demand, and declining growth of productivity in industry and agriculture (see Reynolds 1983: 30). Although some initiatives were introduced in the 1970s to alter existing development policies, rapid growth in the economy masked the urgency of the need for reform. Then, in 1982, a major financial crisis shook Mexicans into a thoroughgoing awareness that major changes were needed. To recover its capacity to grow, the economy would have to become more efficient, and the country would have to become more internationally competitive in the production of goods and services through a strategy of export-oriented growth (see Bennett and Sharpe 1985:227–232; Gil Días 1985; Reynolds, 1983; Weintraub 1984).[1]

1. This concern to reorient basic development strategy toward an export orientation was particularly keen after 1982, when it became painfully apparent that the country could no longer generate the foreign exchange it needed through credits or the profits from petroleum

Table 13. GDP growth rates, 1940–1985 (percent)

1940–1950	6.7
1950–1960	7.0
1960–1969	7.2
1970–1974	6.8
1975	5.6
1976	4.2
1977	3.4
1978	8.3
1979	9.2
1980	8.3
1981	7.9
1982	−0.5
1983	−5.3
1984	3.7
1985	2.7

Sources: 1940–1950, Grindle 1977a:77; 1950–1984, IMF 1985:132–133; 1981–1985, IDB 1986:314.

Mexico's future economic development, however, would depend not only on its capacity to reachieve high rates of growth and to become internationally competitive but also on the country's ability to generate productive jobs for its population, raising incomes and general welfare and creating the conditions for more sustainable growth. I have argued that, because Mexico's major cities are already stretched far beyond their capacities in human terms, the creation of employment opportunities would need to be focused on the country's smaller cities, towns, and rural areas. In this book, we have seen how massive labor migration has been nurtured by the lack of economic opportunities in rural areas. We have been concerned with suggesting ways in which the country's rural and semirural regions, where a large portion of the population continues to live, can become more viable economically, increase their labor-absorptive capacities, and contribute more effectively to the stimulation of both agricultural and industrial development in the country through increased income and demand. We have also considered the politics of reorienting rural development policy away from almost exclusive concern with questions of agricultural production toward broader issues of employment and rural industrialization.

Pursuit of policies and programs to stimulate employment-generating

exports (see Wyman 1983:21–23). By the early 1980s, economists concurred that, "with respect to Mexico's [international] debt, the situation has reached the point where even if Mexico were able to repay its dollar debt obligations to the United States, it could do so in real terms only by exporting goods and services in greatly increased amounts" (Reynolds 1983: 38).

rural development depends fundamentally on policies and programs formulated and carried out in Mexico. Nevertheless, national development strategies and the role of rural areas within them are centrally affected by an international environment that is strongly shaped by the actions of the United States. Issues of U.S.-Mexican relations grow out of the extensive interconnectedness of the two economies and societies and the differences in levels of development between them (see Cornelius and Craig 1984; Erb and Thorup 1984; Musgrave 1985; Ronfeldt, Nehring, and Gándara 1980; Smith 1985). This chapter explores the extent to which national and rural development strategies in Mexico are influenced by economic conditions and policy issues in the United States, with particular attention to issues of trade, immigration, and finance and investment. Then, in the conclusion, issues raised in Chapter 1 are reconsidered from the perspective of the analysis reported in this and previous chapters.

The International Connection: Mexico and the United States

For two thousand miles the United States and Mexico face each other across what has been defined as "the world's most dramatic border between 'North' and 'South'" (Erb and Thorup 1984:vii). In the 1980s, the United States had approximately $5 billion in private investment in Mexico, with 74 percent of it concentrated in the manufacturing sector (Weintraub 1984:168). Trade with Mexico amounted to more than $26 billion a year. Some 300,000 U.S. citizens resided permanently in Mexico, and tourism from the United States sustained a huge industry in Mexico. In the same period, there were some 7 million citizens of Mexican descent living in the United States and an additional 2.5 million Mexican citizens residing in the country. On a more immediate level were factors such as Mexico's enormous reserves of oil and its massive foreign debt which, if left unpaid, could signal the bankruptcy of major banks in the United States. Such conditions underscore the deep economic and social interconnectedness of the two countries.

U.S.-Mexican relations have never been easy; a legacy of conflict, intervention, and Mexican vulnerability to U.S. power ensures that, even in the best of times, Mexico will be suspicious of U.S. intentions, and the United States will fail to comprehend its neighbor's point of view (see especially Smith 1985).[2] Beyond the historical legacy, how-

2. A good example of how Mexican perception of pressure from the United States can encourage the government to make policy decisions that stress independence from the United

ever, there are real divergences of interest between the two countries on issues of trade, energy, migration, and foreign policy. These differences will not disappear through periodic commitments to good neighborliness or rhetorical pronouncements about the mutuality of goals. The differences are real, and their resolution will deeply affect important interests in the United States and Mexico (see Pellicer de Brody 1981).

In the United States, a "Mexico" policy does not exist. Instead, there is "a weltering variety of changing and often contradictory policies made by a multiplicity of separate and uncoordinated departments, agencies, and committees at the federal (executive and congressional branches), state, regional, and local levels" (Bagley 1981:19; see also Wyman 1981). Domestic groups pursue their interests through this multitude of channels and decisionmakers; when discrete policies are finally determined that affect Mexican interests, Mexico as a foreign policy issue is rarely considered (Rico 1981:193). As a corollary to this process, U.S. policy toward Mexico is more frequently decided in the Treasury, Commerce, Energy, or Agriculture departments and related congressional committees than in the State Department or the Office of the Presidency (see Erb 1982; Wyman 1981). The fact that domestic political factions and decisionmaking sites are often at odds over United States–Mexico issues exacerbates the tendency toward contradictory policies and the lack of an overarching Mexico policy (see Mares 1982). Not surprisingly, then, the links between various U.S. policies and the future development of Mexico are generally overlooked.

In Mexico, a more unified perspective relative to foreign policy issues is complemented by a presidentially centered decisionmaking apparatus (see Smith 1985). Mexico's principal concerns in its relations with the United States have been its own economic development and the maintenance of its national sovereignty vis-à-vis the "Colossus of the North." In fact, national sovereignty, a vital ingredient of Mexican nationalism, is defined primarily in terms of Mexican political and economic independence from the United States. For Mexico, interdependence with the United States has historically been interpreted in terms of U.S. hegemony; for most Mexicans, the country's capacity to develop is fundamentally constrained by the presence of its rich northern neighbor (Story 1982; see Hansen 1979:27 for a discussion). Generalized perspectives about differences in power between the countries color the

States is the GATT decision of 1980. Although domestic political pressures figured in the decision not to join the GATT, an important factor influencing Mexico's stance was the clear preference of the United States for a multilateral approach to trade between the two countries. See Story (1982) for a detailed discussion of the politics of the GATT decision.

specific issues that arise between them. Thus "what may appear to be narrow technical matters to Americans, such as the price of a product or other terms of a contract, may be viewed among Mexicans as sensitive political issues having profound implications for their country's future" (Ronfeldt, Nehring, and Gándara 1980:vii; see also Bagley 1981:16–19). Although many argue that Mexicans must overcome their fear and paranoia about the United States and be both more assertive and more reciprocal in negotiations, the legacy of the past is strong (Erb 1982; Weintraub 1984).[3]

Perceptions of U.S. dominance signal Mexican appreciation of the country's vulnerability to changes in the international economy and an understanding of the role of the United States in defining those conditions. Although more unified in its capacity to make decisions, the Mexican government does in fact have a limited range of options because of its dependence on international markets for oil and agricultural products and its massive foreign debt.[4] Given the structures of U.S. decisionmaking concerning Mexico, and given Mexico's sensitivity to the power of the United States—not to mention the country's frequent experience that it can expect little in terms of partnership from the United States—the potential for working out a relationship more attuned to Mexico's development needs is fraught with hazards. This is particularly true with regard to issues of trade, immigration, and finance and investment.

Trade

In 1980 more than 65 percent of the total value of Mexico's merchandise imports came from the United States (see Table 14). Equally important for Mexico's development, if not more so, the United States received more than 64 percent of the country's merchandise exports (see

3. Those who argue for greater assertiveness and greater efforts toward reciprocity refer to what they interpret as Mexico's avoidance of negotiating opportunities. They suggest that Mexico should develop a greater capacity to "play the lobbying game" in Washington, D.C., in order to make its interests known in both legislative and executive branches. At the same time, they indicate that Mexico is generally unwilling to make concessions during serious negotiations that would make for a more "giving" attitude on the part of the United States. (This summary is based on interviews with officials in Washington, D.C., December 1985, and Erb 1982.)

4. Thus Cornelius and Craig report, "for every $1 drop in the price of a barrel of oil, Mexico loses $500 million in revenues. Mexico is equally vulnerable to changes in interest rates prevailing in the United States and world money markets. For every 1 percent rise or decline in such interest rates, Mexico will have to pay or will save nearly $700 million in interest payments on its foreign debt" (1984:424).

Table 14. Mexico's merchandise imports, by country and area, 1976–1980

Country	1976	1977	1978	1979[a]	1980[a]
			Millions of dollars		
United States	3,765	3,485	5,023	7,637	8,638
Canada	141	166	162	197	262
Europe[b]	1,275	1,131	1,768	2,508	2,461
South America	245	244	346	576	483
Caribbean and Central America	163	73	64	123	170
Other[c]	441	390	780	1,056	1,161
Israel	0.5	0.5	1	4	4
Japan	306	295	590	787	693
People's Republic of China	9	9	24	43	48
Total	6,030	5,489	8,143	12,097	13,175
			Percentage		
United States	62.4	63.5	61.7	63.1	65.6
Canada	2.3	3.0	2.0	1.6	2.0
Europe[b]	21.1	20.6	21.7	20.7	18.7
South America	4.1	4.4	4.2	4.8	3.7
Caribbean and Central America	2.7	1.3	0.8	1.0	1.3
Other	7.3	7.1	9.6	8.7	8.8
Israel	—	—	—	—	—
Japan	5.0	5.3	7.2	6.5	5.2
People's Republic of China	—	—	—	—	—
Total	100.0	100.0	100.0	100.0	100.0

Note: A dash means less than 0.5 percent.
[a]Preliminary figures. Figures for 1980 are for January through September.
[b]Predominantly Western Europe.
[c]In the years shown, Japan generally has accounted for 75 percent of the "other" category.
Source: Weintraub 1984:70.

Table 15). Mexico supplies about 5 percent of all U.S. imports and receives about 7 percent of its exports (Weintraub 1984:34). It provides the largest developing country market for U.S. exports, accounting for 12.6 percent of exports to all Third World countries (see Sewell, Feinberg, and Kallab 1985:182). The interrelationship of trade between the two countries is impressive and manifests itself in a variety of ways. Technology transfers from the United States, for example, have a direct impact on Mexico's process of industrialization (see Thorup 1986; Whiting 1984). In agriculture, U.S. markets are vital to the labor-intensive production of fruits and vegetables that has become so important in northern and central Mexico. In the 1980s, Mexico supplied about 60 percent of U.S. imports of winter fruits and vegetables (Mares 1982:81). Historically, these agricultural products have been subject to seemingly arbitrary regulations about quality and packaging as well as the object of antidumping complaints by producers in the United

Table 15. Mexico's merchandise exports, by country and area, 1976–1980

Country	1976	1977	1978[a]	1979[a]	1980[a]
			Millions of dollars		
United States	1,854	2,399	4,411	6,147	7,028
Canada	48	44	61	74	96
Europe[b]	367	397	572	1,098	1,716
South America	315	371	381	419	444
Caribbean and Central America	191	215	240	271	466
Other	214	224	552	905	1,203
Israel	65	70	106	299	432
Japan	100	82	171	248	333
People's Republic of China	9	17	123	129	58
Total	2,989	3,650	6,217	8,914	10,954
			Percentage		
United States	62.0	65.7	70.9	68.9	64.2
Canada	1.6	1.2	1.0	0.8	0.9
Europe[b]	12.3	10.9	9.2	12.3	15.7
South America	10.5	10.2	6.1	4.7	4.0
Caribbean and Central America	6.4	5.9	3.9	3.0	4.2
Other	7.2	6.1	8.9	10.2	11.0
Israel	2.2	1.9	1.7	3.3	3.9
Japan	3.3	2.2	2.8	2.8	3.0
People's Republic of China	0.3	0.4	2.0	1.4	0.5
Total	100.0	100.0	100.0	100.0	100.0

[a]Preliminary figures. Figures for 1980 are for January through September.
[b]Predominantly Western Europe.
Source: Weintraub 1984:71.

States.[5] The tomato wars of 1969 and 1978 are well-documented cases of the dynamics and problems of U.S.-Mexican agricultural trade (see Mares 1982; Weintraub 1984:48–53). In addition, the U.S. market has for many years been the source of grain imports that Mexico has needed to make up for domestic supply shortfalls (see Sanderson 1983).

Energy is an obviously critical trade issue between the two countries, as is the openness of the Mexican market to U.S. products—traditionally, Mexico's market has been more closed to the United States than vice versa. The mutual harm that can occur through trade declines is equally impressive: in 1982, U.S. trade with Mexico declined by 32 percent and as a result an estimated 250,000 jobs were lost in the United States (Cornelius and Craig 1984:422). The economic crisis in Mexico is estimated to have cost the United States some $10 billion in exports in

5. Health and sanitary regulations, tariff structures, policies about chemical residues, and size and grade requirements, for example, have been altered at the behest of U.S. producer groups concerned about competition from Mexican agricultural products (see Bagley 1981: 19).

Table 16. Nonagricultural wage and salary employment and percentage unemployed in Texas border standard metropolitan statistical areas, January 1982 and January 1983

Item	Brownsville	McAllen	Laredo	El Paso	Texas
Employment					
1982	65,700	83,150	37,500	170,400	6,271,700
1983	58,250	79,300	30,450	162,900	6,168,700
Percentage decline	11.3	4.6	18.8	4.4	1.6
Unemployment rate					
1982	11.4	14.0	11.0	9.2	5.9
1983	17.7	20.5	27.3	13.3	8.5

Source: Hansen 1985:8.

the two years between 1981 and 1983 (see Weintraub 1985:99). The interconnectedness of the economies is perhaps most evident in the border region. Table 16 demonstrates the damaging impact of the Mexican crisis of 1982 on the economies of border cities in Texas, cities that depend heavily on trade with Mexico.

Trade relations with the United States are central to Mexico's ability to create more productive sources of employment within its borders. The country can of course attempt to diversify the destination of its exports, as it has been doing in recent years, but given the massive nature of trade dependence on the United States, major changes in trading partners are unlikely in the near future, and it is difficult to imagine that U.S.-Mexico trade relationships will become any less important to either country. Instead, Mexico's development policy, and its ability to employ increasing numbers of people in productive jobs, depend in no small part on increasing access to U.S. markets. Yet at the very time when such markets became more vital to Mexico, protectionist trends in the United States made them less accessible to the country's exports in the 1980s (see ODC 1985).

In spite of the extensiveness and importance of the trade relationships between Mexico and the United States, there have been no general trade agreements to provide a stable framework for discussing trade issues between the two countries. Instead, issues have been decided on a case-by-case basis, and conflicts over countervailing duties, unfair trade practices, and "dumping" overwhelmed the agenda of discussions at this level. In part, this situation resulted from differing perspectives held by the Mexican and U.S. governments on the role of international trade in economic development. Mexico has considered trade policy to be central to a strategy for national development, whereas the United

States has been committed to establishing and maintaining generalized international agreements about the rules that govern trade (see especially Bagley 1981; Weintraub 1984).

Thus Mexican export policy has been fueled by a plethora of incentives to reward exporters, especially exporters of manufactured goods. Credit, energy subsidies, tax incentives, foreign investment regulations, and rebates have been used to spur the industrialization that has been at the core of the country's development strategy. The policy tools chosen by the government allowed it considerable freedom to reward and punish individual manufacturers or importers for contributing or not to national development goals. The legacy of this policy framework will be difficult to dispel. Nevertheless in the period of the late 1970s to the mid-1980s, export promotion of industrial goods was accompanied by a shift toward a deprotectionist policy for goods manufactured for the domestic market (Bennett and Sharpe 1985:227–232). This largely took the form of lowering Mexico's high tariff walls on selected imports, domestic content requirements, and a considerable winnowing of the number of items subject to import quotas.[6] Then, in November of 1985, under increasing pressure from the United States, the government announced its intention of seeking membership in the General Agreement on Tariffs and Trade (GATT), a move that it had rejected in 1980.

The United States has taken a different approach to international trade, one based on concern about the fairness of the rules of the game for engaging in trade. For the United States, the incentives and subsidies that have been the basis of Mexican industrialization policy raise the specter of "unfair trade" (see Samet and Hufbauer 1982 for a discussion). Not surprisingly, then, in a context lacking a general framework of principles to regulate trade between the two countries, the issues that arise consistently have been disputes over quotas, "unfair" subsidies, countervailing duties, and tariffs. Complaints from U.S. manufacturers were particularly notable in the mid-1980s, when a poor international situation for U.S. manufactures heightened their concern about foreign competition (Erb and Thorup 1984:16). At the same time, Mexico's economy, devastated by the end of the oil boom of the late 1970s and

6. Mexican policy began to shift, encouraging greater international competitiveness from its domestic entrepreneurs in the mid-1970s, when barriers on selected imports were lowered. Between 1975 and 1980, about five thousand items, 35 percent of the value of Mexican imports, were removed from the list of items requiring import licenses. Similarly, quotas on imports were relaxed, and some tariffs were reduced (see Hufbauer, Smith, and Vukmanic 1981:142; Weintraub 1984:48–53).

massive international debt obligations, made the government particularly eager to pursue foreign markets for the country's products (see del Castillo 1985).

The United States would prefer to deal with United States–Mexico trade issues within the multilateral context of the General Agreement on Tariffs and Trade. Mexico, until late 1985, refused to join the GATT, although it participated in the Tokyo round of global trade negotiations in the 1970s and had negotiated an agreement to join in 1979. At that time, however, participation in the GATT represented for Mexico the acceptance of severe constraints on its use of national development policy tools such as subsidies, quotas, local content requirements, and export incentives, even while it would become eligible for less arbitrary procedures in resolving trade disputes with the United States and other countries (see del Castillo 1985; Hufbauer, Smith, and Vukmanic 1981: 142). Mexico wanted instead a "special relationship" in its trade with the United States, a bilateral approach that the United States has refused to accept as an overall framework for negotiation. The resulting standoff meant considerable instability and even arbitrariness when specific Mexican products were considered. Because it was not a formal participant in the GATT, Mexico was unable to appeal allegations of unfair trade injuries by U.S. manufacturers. Thus it had little protection against the imposition of countervailing duties, which, through the Trade Agreement Act of 1979, could be assessed and imposed rapidly.[7] Until the mid-1980s, a two-tier system, in which some items were eligible for import into the United States through the Generalized System of Preferences (GSP) and other items are treated as if they were part of the GATT system, left most trade issues to be resolved through unilateral decisions by the U.S. (del Castillo 1985:24; see ODC 1984c for a discussion).[8]

7. The Trade Agreements Act of 1979 made it mandatory to utilize "fast track" methods for investigating charges of unfair trade practices brought by U.S. interests. Hofbauer, Smith, and Vukmanic note, "Under these fast track procedures, decisions are made under severe time constraints on the best evidence available—a particular problem for respondent importers and foreign producers. Further, the antidumping and countervailing duty administrative procedures were modified to allow new opportunities for judicial review of intermediate and final decisions made by the Commerce Department and the International Trade Commission. Finally, the secretary of commerce lost the authority that the secretary of treasury had to waive countervailing duties" (1981:144).

8. The Trade Act of 1974 created the Generalized System of Preferences in order to give preferential treatment to goods from developing countries. In general, the GSP has benefited more advanced developing countries like Mexico than it has poorer countries (see ODC 1984c). About 20 percent of Mexico's merchandise exports to the U.S. was covered by the

The problems of working out agreements on trade are formidable. They are perhaps greatest in the United States, where a highly decentralized and fragmented decisionmaking process is further fractionated by a multitude of highly mobilized groups with very specific stakes in any general process of negotiation about trade issues. The current system of case-by-case decisionmaking has avoided dealing head-on with these difficult issues in which mutually conflicting interests exist. Maintenance of the system, however, would leave U.S.-Mexican trade in disarray, making it difficult for either country to pursue its goals. The long-term solution to Mexico's major problems of development—such as lack of employment opportunities and a massive foreign debt—lies in the country's ability to produce competitively for domestic and international markets and to have regularized access to those markets. Despite the tensions and disagreements in perspectives indicated above, trade between the two countries is likely to increase rather than decrease. Without a more generalized commercial agreement between them, however, there is little certainty for potential investors and added risk of failure for those who are committed to improving the employment situation in Mexico.

Immigration

The history of U.S. immigration policy toward Mexico is replete with examples of bilateral agreements that have served interests on both sides of the border. The policy legacy is also characterized by numerous unilateral actions by the United States to halt immigration whenever a broad domestic consensus on the need for Mexican workers in the economy ceases to exist. The Bracero Program (1942–1964) was one special bilateral arrangement for using Mexican workers to meet a U.S. labor shortage. In 1986, a guest agricultural worker program modeled on the European experience was authorized by Congress. Such special programs have not been initiated from concern about Mexico's de-

---

GSP in the mid-1980s, and Mexico indicated considerable unease about increased discretionary powers in the U.S. executive branch to remove items from the GSP list under the Omnibus Tariff and Trade Act of 1984, by which such items were made subject to "unfair trade" procedures (del Castillo 1985:12). The country was quick to believe that U.S. actions to remove items from the list were political acts directed against its national development goals (del Castillo 1985:15, 18; Erb and Thorup 1984:14–15). Del Castillo found that, "between April 1, 1983 and March 31, 1984, a total of 97 products were removed from GSP listing, and of these, 63 are of Mexican origin. That is, 90.2% represent Mexican exclusions" (see del Castillo 1985:15, 18). For its part, the United States considered that Mexico, as a newly industrializing country, should "graduate" to a less protected status in its trade relations and thus pushed hard to persuade its neighbor to join the GATT (see ODC 1984c for a discussion). The process of graduation, initiated in 1981, meant that export items worth $134 million from Mexico were removed from the GSP in 1984 (Erb and Thorup 1984:14–15).

velopment problems and have been primarily concerned with supplying a steady and stable source of low-wage unskilled labor for agriculture. Nevertheless, they have satisfied some of the unmet need for jobs within Mexico. In contrast, restrictive U.S. immigration policy has tended to create problems for both Mexico and the United States by ignoring the special conditions created by proximity, a highly permeable border zone, and differences in levels of development between the two countries.

The Immigration Reform and Control Act of 1986 was the most extensive immigration reform effort initiated in the United States since 1952 and resulted from a compromise worked out among a variety of labor, business, local government, Hispanic, population, environmental, and civil rights groups.[9] Its central provisions sought to discourage the employment of illegal aliens through sanctions against employers who hire undocumented workers (see Cornelius 1983:142). The logic of the bill was clear: if employment opportunities are eliminated, then immigration can be curbed. There is some evidence suggesting that this assumption should be questioned. Legislative restrictions on immigration are extremely difficult to impose where there exists a two-thousand-mile border that is nearly impossible to police effectively short of militarization. Moreover, in this book, we have reviewed evidence supporting the assertion that the labor market, not legal strictures, strongly determines the flow of immigration. As long as economic incentives to migrate outweigh incentives to remain in the local area, and as long as individual households are able to choose migration as an alternative, it is reasonable to assume that they will do so, regardless of the risk factors or penalties involved. As we have seen, rural households have proved adept at minimizing the risk of migration and at adjusting to the loss of household labor during migratory periods. Economic crises in Mexico in the 1970s and 1980s actually increased wage differentials and therefore increased the incentive to migrate. Moreover, where demand for workers exists extensively, employers are likely to find creative ways to circumvent the hiring sanctions that might be legislated.[10] It should also

9. It is estimated that only about 15 percent of illegal aliens are agricultural workers. For coverage of debate on the bill, see U.S. Congress, *Daily Digest* (September 11–13, 16–18, 1985). See also Teitelbaum (1985: chap. 5) for a discussion of U.S. immigration policy.

10. "Not a single person has even been convicted under California's employer sanctions law since its passage in 1971, and nationwide, state-level employer sanctions laws have resulted in only five convictions. . . . Similarly, the major General Accounting Office report . . . concluded that in all of the 20 countries surveyed by the GAO, 'laws penalizing employers of illegal aliens were not an effective deterrent to . . . illegal employment. . . . Employers either were able to evade responsibility for illegal employment or, once apprehended, were penalized too little to

be remembered that, according to many analysts, the United States will be facing a labor shortage by the 1990s, particularly in low-wage service, manufacturing, and agricultural jobs (Cornelius 1981a:9; Cross and Sandos 1981:49–50; Reynolds 1979:139). If this is the case, jobs in the United States will exert a powerful pull that will not easily be broken by legal strictures.[11]

In this book, I have attempted to demonstrate that the root of high illegal migration to the United States is the failure of Mexico's development to create productive sources of employment, particularly for its rural population. Individuals migrate from poor rural areas because of the lack of economic opportunities in their communities. They are drawn to the United States because of higher wages and a tradition of migration to U.S. destinations. The solution to high levels of illegal immigration, then, lies not in the United States but in the pursuit of an employment-generating growth strategy for Mexico, one that is particularly attuned to the problems and needs of rural areas. Restrictive immigration policy is likely to have little impact on the course of Mexico's development because the dynamics involved in migration are economic, not legal.

In previous chapters, we discussed the importance of remittance income for sustaining families in rural areas in Mexico and for improving general conditions of welfare. This important side effect of migration became particularly notable in the 1970s and 1980s when devaluations of the Mexican peso increased the logic of remitting dollars to families remaining behind rather than attempting to support them in the United States. In Chapter 4, several examples were given of ways in which remittance income had been invested productively in four rural regions. Depressed conditions in the 1980s encouraged the migrants themselves to think more seriously about the contributions that migration can make to their families and communities. No longer was immi-

---

deter such acts'. . . . Nothing that the U.S. government has done in the last 100 years has appreciably reduced the demand for Mexican and other foreign labor in our economy" (Cornelius 1983:142–143, 148; see also Portes 1982; Zolberg 1982). The GAO reversed its stand on this issue in 1985 (see GAO 1985).

11. A large body of evidence about Mexican immigrants, however, indicates that: (1) there is, in general, little overlap between the kinds of jobs held by illegal immigrants and those sought by U.S. workers, although localized competition has been noted; (2) particular industries, such as tourism, agriculture, and small-scale manufacturing, would be in difficult straits without immigrant labor; (3) undocumented migrants generally contribute more to local and national coffers through taxes on income and consumption than they cost in terms of local or social services; and (4) undocumented aliens do not contribute to a rise in crime rates (see Craig 1981:15–16).

gration to the United States considered by large numbers of Mexican youth to be simply "an adventure" or even a "rite of passage" to more serious adulthood. Instead, it took on a very serious purpose for families who found little possibility for sustaining themselves without it.[12] Thus for many reasons migration and the remittance income it generates need to be viewed much more as positive sources of financing for the development of Mexico's depressed rural areas. Finally, temporary migration to the United States adds to the skills of workers, who are then able to return to their communities with greater potential to contribute to local development initiatives. Chapter 4 described several examples of this human capital formation.

The message for policymakers from these observations can be clear: immigration to the United States should be evaluated in terms not only of its impact on the United States but also of its potential to encourage the long-term development of Mexico's rural regions. In fact, immigration policy is an issue area where there may actually be less conflict of interest and more mutuality of goals than is generally thought. U.S. labor needs may require it, and Mexican development can benefit from it. Some have argued, of course, that restrictive immigration practices serve as a kind of "shock treatment" for the Mexican government. In this view, from fear of the social unrest that will ensue if the U.S. border is closed, the Mexican government will reorient its development policy and will begin to accord rural areas the attention they deserve through a more equitable and redistributive approach to development that creates jobs (see Cornelius 1981c for a discussion). The logic of this argument is persuasive; governments often show considerable capacity to accomplish the unexpected—and even the impossible—when the fear of social unrest is great enough. Unfortunately, a shock treatment, even if it were feasible to close the border, is unlikely to generate this response in Mexico. Simply stated, the Mexican government does not have the financial and administrative capacity to respond to employment needs in the countryside in the short or medium term. However dedicated it is to the generation of jobs, it will not be able to meet the need except over a considerable period. Moreover, the threat of social unrest is a threat not just to the Mexican political system but also to fundamental U.S. interests in a stable and secure border.

12. Some indication of the impact of Mexico's economic crisis on migration to the United States is that apprehensions of undocumented workers by the U.S. along the Mexico-U.S. border increased by 40 percent in 1983 over 1982. An additional 40 percent increase was reported by the U.S. Immigration and Naturalization Service in 1986 (see *LAM* 1986:268).

Finance and Investment

Although Mexico's economic crisis of the 1980s reflected the situation of developing countries within a deeply distorted international system, domestic development policies of the 1970s and 1980s also undeniably contributed appreciably to its problems of debt, trade, and investment. Extensive foreign borrowing and massive government expenditures beginning in the early 1970s helped create an external debt of 96 billion dollars by 1984, the second highest foreign debt in the world (*LAM* 1985:232, 1986:51; see Table 17). In that year, the debt equaled 55.5 percent of the country's gross domestic product, and $10 billion was required to service it. This $10 billion accounted for 71 percent of the country's trade surplus and 40 percent of its total export revenue. The debt was expected to rise to more than $115 billion by 1990 (*LAM* 1986:53). Clearly Mexico had banked on its petroleum to steer clear of such massive foreign indebtedness; falling oil revenues in the 1980s greatly exacerbated its problems. A debt of this magnitude was a major impediment to Mexico's capacity to reachieve high rates of growth and to take the actions necessary to create jobs for its population.

Unfortunately, many of the short-term adjustment policies Mexico adopted after 1982 worked against the country's capacity to build a more dynamic economy in the longer term. In November of 1982, the country signed a debt renegotiation agreement with the International Monetary Fund; in the summer of that year, Mexico ran out of the foreign reserves needed to service its massive debt (see Wyman 1983 for a discussion). Under the circumstances, the agreement was a reasonable one from Mexico's perspective, and in return the government agreed to

Table 17. Total external debt, Mexico, 1977–1985 (U.S. $ billions)

1977	29.7
1978	33.9
1979	40.2
1980	50.7
1981	74.9
1982	80.0
1983	86.0
1984	95.9
1985	98.0
1988[a]	110.0
1990[a]	115.4

[a]Estimated.
Source: Trebat 1985:40; *LAM* 1985:243, 232; *LAM* 1986:51.

a program of fiscal austerity to deal with the mounting rate of inflation, to secure foreign exchange activities, and to pay the interest on the debt (Wyman 1983:13). The public sector deficit was to be drastically reduced, and the level of public sector borrowing from abroad was restrained. In the years between 1982 and 1985, Mexico was lauded for its success in meeting IMF standards and in responding positively with its austerity program.

Mexico's austerity measures, however, were pursued at great cost. Their effect was to reduce real income significantly and to limit the rate of investment in the economy so that few new jobs were created and unemployment mounted. Between 1982 and 1985, real wages dropped by 50 percent. Although the middle class may have been the group to see its income drop most, the poor, on the margins of subsistence to begin with, suffered most significantly in terms of increasing pressure on their ability to meet household subsistence needs. Labor migration, always an attractive possibility, became a virtual necessity for many, and the rationality of migrating to the United States increased substantially. Although inflation continued to be high, successive peso devaluations meant that the rising price of the dollar more than offset increased prices in Mexico; dollar remittances often made the difference between starvation and survival for many Mexican households (see Conroy, Coria Salas, and Vila González 1979:34; *LAM* 1985:229).

The responses to the economic crisis have dampened the possibilities for Mexican recovery by discouraging investment and severely limiting the productive capacity of the economy. New investments were not made. Existing industrial capacity remained underutilized because of lack of capital and a highly restricted market. In such a situation, incentives to save or to invest were missing. In 1984, the payment of the debt accounted for about a quarter of domestic savings, and capital flight continued apace. As we have seen, after 1982, the Mexican government made significant efforts to control public spending, reduce imports, and bring inflation down. Trade restrictions were liberalized, and the country agreed to join the GATT. Moreover, by the mid-1980s, Mexico was demonstrating a much more accommodating attitude toward foreign investment than it had during the previous decade. Such measures were intended to bring greater confidence in the country's economy and thus encourage savings and investment. That the country was able to maintain its tradition of political stability throughout the difficult period of severe economic crisis was a major achievement. Nevertheless, many of Mexico's economic problems—and therefore its capacity to recover—reflect international conditions over which it had no control. Lowered interest rates, better commodity prices, and opportunities for trade are critical to the country's future yet are beyond its capacity to influence. Achieving these goals depends in no small part on U.S. domestic policies for cutting its own public deficit, bringing down interest rates, avoiding recession, and resisting trade protectionism (ODC 1984a:b; Sewell, Feinberg, and Kallab 1985). In the case of the issues of investment and finance, then, policies that would encourage a more supportive international economic environment for Mexico are not on a bilateral agenda between the two countries. They are first on a domestic U.S. agenda and second on a multilateral agenda for establishing a dynamic international economy and helping a range of developing countries deal with the problems of debt and development more effectively.

The Future of Rural Development

This book began with a concern for understanding the causes and consequences of temporary labor migration from rural areas. Increasingly, such migration is a strategy adopted by rural households throughout the Third World when they are squeezed by expanding population pressure on the land, a declining resource base in agriculture, increasing landlessness, discriminatory government policies, and industrialization strategies that fail to generate sufficient employment in urban areas. Temporary migration differs in its characteristics from the permanent rural-to-urban movements that have been the focus of most previous considerations of migration: its purpose is generally to maintain a family in a rural area rather than to find an alternative to rural life; it is often repeated a number of times in the life of a migrant; and through remittance income it is an important way in which capital is infused into rural areas from urban and international sources. This study focused on the factors in Mexico's rural areas that push migrants away from their communities in search of work but that also pull them back again with regularity. These push and pull factors are important for understanding constraints on agricultural development in the country.

This book also stressed the extent to which rural areas have become dependent, not on agriculture or on a diversified local economy, but on infusions of resources from nonlocal sources. Many discussions of rural poverty and underproductivity assume that economic underdevelopment and dependence exist because rural areas are isolated and remote from dynamic modern sectors of the national and international economies. The case of Mexico, however, suggests a very different picture of rural underdevelopment. Most temporary labor migration in the country originates in the countryside; the regions responsible for the majority of labor migrants are the most densely populated rural areas of the country, depend the most on scarce and often erratic rainfall for agricultural pursuits, and show the greatest fragmentation of landholding and landlessness. They are also the rural regions that have been most fully integrated into national and international economies and are most penetrated by political, social, and economic systems of the larger society. Thus they are not the areas where the most traditional forms of agriculture predominate but are instead areas that have been strongly affected by market forces and government development policies.

The research presented here also suggested that rural communities effectively exploit nonlocal opportunities for employment. In general, however, such activities have led only to the maintenance of a meager subsistence and a continued dependence on the availability of employment elsewhere. Thus although temporary labor migration is a rational choice for rural households to make, its economic causes and consequences indicate broad problems of underdevelopment and public policy. An initial concern with temporary labor migration helped focus attention on the fact that agricultural pursuits are only one among many income-generating activities that sustain rural households and communities. Most studies of rural areas and their development have taken as axiomatic that rural households have access to land and that they generate the major portion of their income from the land. Increasingly this view distorts what occurs in vast numbers of rural communities throughout the Third World. Landlessness and declining employment opportunities in agriculture are increasingly marked (see, for example, Lassen 1980). Nonfarm employment, much of it through temporary labor migration, makes it possible for rural inhabitants to maintain their homes in rural areas. Off-farm employment cannot be considered a residual source of income for rural households; for many, it is their principal means for meeting their economic needs. The importance of nonfarm income appears to be increasing with time, and such employment generally represents a search for alternatives to poor prospects in agriculture rather than an outcome of modernization of farm production. This pattern contrasts significantly with the experience of agriculture-led rural development that occurred in the "success stories" reviewed in Chapter 1.

In a more general sense, the perspective presented in this book is that agricultural development responds to only part of the problem of rural poverty and welfare because many rural areas are so resource poor that modernization of agricultural production becomes an extremely costly and even impractical alternative. A broad-based agriculture-led development strategy requires much more equitable access to land and other resources than is the case in Mexico and many other countries. Where development strategies of the past have created a politically and economically significant sector of large-scale modern commercial farmers, however, the prospects for altering fundamental allocations of agricultural resources may be remote. In such cases, agriculture may not have the capacity to serve as an effective "engine" for rural development. The issue for rural development policy is whether rural communities can be linked more effectively to regional and urban activities so that employment opportunities are generated even where agricultural modernization is not a viable option. If such an approach is adopted, employment generation should become the central criterion by which policies, programs, and projects to stimulate rural development are assessed and is the measure applied to private and public investments.

The analysis also indicated that there are no easy solutions to the dilemmas posed by the legacies of Mexico's past development. Many factors affecting the country's ability to grow and respond to severe problems of inequity and poverty are shaped by international economic conditions and policies over which it has little influence. Its national political system has demonstrated considerable capacity to make difficult economic choices while also maintaining its historical tradition of social and political stability. At the same time, the very strength of the institutions that have made these situations possible places limits on the potential for poor rural inhabitants to become more active participants in national policymaking, able to assert their demands for a more equitable development strategy in the country. This is particularly true in terms of the economic and political obstacles to pursuing the broad-based agriculture-led development strategy advocated by many rural and national development specialists.

Instead, Mexican development policymaking forces analysts to consider not just the question of what ought to be but also the issue of what can be, given a realistic appreciation of the characteristics of a country's political economy. At the level of the household, the community, the município, the national political and economic systems, and the international arena, there are no quick fixes for Mexico. Instead, there are some possibilities for altering important policies and initiating regional and local activities that may, over the long term, integrate rural Mexico more fully into the country's development process and make it more able to contribute to and benefit from that process. Ultimately, even long-term responses will result only from the actions of individuals and institutions at local, national, and international levels who believe that it is possible to change a deplorable reality into a more acceptable future. Rural families in Mexico have long demonstrated their resourcefulness and commitment to resolving the problems they face. It remains for them and for others to demonstrate that the same kind of resourcefulness and commitment can be generated to address the community, national, and international consequences of rural underdevelopment.

Bibliography

Adams, Dale W. 1983. "Mobilizing Household Savings through Rural Financial Markets." In J. D. Von Pischke, Dale W. Adams, and Gordon Donald, eds., *Rural Financial Markets in Developing Countries*. Baltimore, Md.: Johns Hopkins University Press.

Alarcon, Francisco. 1982. "Family Planning Activities in Mexico." In Jorge Martínez Manautou, ed., *The Demographic Revolution in Mexico*. Mexico City: IMSS.

Alba, Francisco. 1978. "Mexico's International Migration as a Manifestation of its Development Pattern." *International Migration Review* 12:4 (Winter), 502–513.

Alejo, Francisco Javier. 1983. "Demographic Patterns and Labor Market Trends in Mexico." In Donald L. Wyman, ed., *Mexico's Economic Crisis: Challenges and Opportunities*. Monograph No. 12. La Jolla: University of California, San Diego, Center for U.S.-Mexican Studies.

Ali, Sayed Ashraf, et al. 1981. *Labor Migration from Bangladesh to the Middle East*. Staff Working Paper No. 454. Washington, D.C.: World Bank.

Alonso, William. 1984. "Recent Developments in Mexico's Decentralization and Urban Policies." Report on urban decentralization in Mexico. Washington, D.C.: World Bank.

Anderson, Bo, and James Cockcroft. 1972. "Control and Cooptation in Mexican Politics." In James Cockcroft, André Gunder Frank, and Dale L. Johnson, *Dependence and Underdevelopment: Latin America's Political Economy*. Garden City, N.Y.: Anchor Books.

Anderson, Dennis, and Mark Leiserson. 1980. "Rural Nonfarm Employment in Developing Countries." *Economic Development and Cultural Change* 28:2 (January), 227–248.

Anzaldúa Montoya, Ricardo, and Wayne A. Cornelius. 1983. *The Report of the U.S. Select Commission on Immigration and Refugee Policy: A Critical Analysis*.

Research Report No. 32. La Jolla: University of California, San Diego, Center for U.S.-Mexican Studies.

Arizpe, Lourdes. 1982. "The Rural Exodus in Mexico and Mexican Migration to the United States." In Peter J. Brown and Henry Shue, eds., *The Border That Joins*. Totowa, N.J.: Rowman and Littlefield.

Astorga Lima, Enrique. 1985. *Mercado de trabajo rural en México: La mercancía humana*. Mexico City: Ediciones Era.

Austin, James E. 1981. *Agroindustrial Project Analysis*. Baltimore, Md.: Johns Hopkins University Press.

Austin, James E., and Gustavo Esteva. 1985. "SAM Is Dead—Long Live SAM." *Food Policy* 10:2 (May), 123–136.

Ayres, Robert L. 1981. "The Future of the Relationship." In Susan Kaufman Purcell, ed., *Mexico-United States Relations*. New York: Academy of Political Science.

Babbitt, Bruce. 1985. "Babbitt Calls for Halving of Mexican Debt Service." Remarks on the Mexican debt crisis presented to the Overseas Development Council, Washington, D.C., November 13.

Bachrach, Peter, and Morton S. Baratz. 1970. *Power and Poverty: Theory and Practice*. New York: Oxford University Press.

Baer, Katherine. 1985. "Export Subsidies in the Mexican Leather Industry: A Thorn in the Side of U.S.-Mexican Relations." M.A. thesis, Law and Diplomacy, Tufts University.

Bagley, Bruce M. 1981. "A United States Perspective." In Susan Kaufman Purcell, ed., *Mexico-United States Relations*. New York: Academy of Political Science.

Bailey, Norman, R. David Luft, and Roger W. Robinson, Jr. 1983. "Exchange Participation Notes: An Approach to the International Financial Crisis: An Opportunity for Constructive Action." Washington, D.C.: Georgetown University.

Barrientos, Guido A., Harmon M. Hosch, G. William Lucker, and Adolfo J. Alvarez. 1984. *Employment Opportunities in Mexico for Undocumented Border Crossers as a Function of Manpower Needs, Part I*. Report prepared for the U.S. Bureau of Prisons, La Tuna, N.M.

Bassoco, Luz María, and Roger D. Norton. 1983. "A Quantitative Framework for Agricultural Policies." In Roger D. Norton and Leopoldo Solís M., eds., *The Book of CHAC: Programming Studies for Mexican Agriculture*. Baltimore, Md.: Johns Hopkins University Press.

Bennett, Douglas, and Kenneth Sharpe. 1980. "The State as Banker and Entrepreneur: The Last-Resort Character of the Mexican State's Economic Intervention, 1917–76." *Comparative Politics* 12:2 (January), 165–189.

———. 1985. *Transnational Corporations Versus the State: The Political Economy of the Mexican Auto Industry*. Princeton, N.J.: Princeton University Press.

Berg, Ronald H. 1985. "The Contribution of Return Migration to the Growth of Rural Capitalism in Peru." Notre Dame, Ind.: University of Notre Dame.

Berry, R. Albert. 1984. *Research Priorities for Employment and Enterprise Development in Rural Regions*. Development Discussion Paper No. 170. Cambridge, Mass.: Harvard University, Harvard Institute for International Development.

Berry, R. Albert, and R. H. Sabot. 1984. "Unemployment and Economic Development." *Economic Development and Cultural Change* 33:1 (October), 99–116.

BLA. Various years. *Business Latin America*, various issues.

Brading, David. 1977. "Hacienda Profits and Tenant Farming in the Mexican Bajío,
1700–1860." In Kenneth Duncan and Ian Rutledge, eds., *Land and Labour in
Latin America*. Cambridge: Cambridge University Press.

Bradshaw, Benjamin, and W. Parker Frisbie. 1983. "Potential Labor Force Supply
and Replacement in Mexico and the States of the Mexican Cession and Texas:
1980–2000." *International Migration Review* 17:3 (Fall), 394–409.

Bruton, Henry J. 1974. "Economic Development and Labor Use: A Review." In
Edgar O. Edwards, ed., *Employment in Developing Nations*. New York: Ford
Foundation.

Burbach, Roger, and Patricia Flynn. 1980. *Agribusiness in the Americas*. New
York: Monthly Review Press.

Bustamante, Jorge. 1983. "The Mexicans Are Coming: From Ideology to Labor
Relations." *International Migration Review* 17:2 (Summer), 323–341.

Byerlee, Derek, Carl K. Eicher, Carl Liedholm, and Dunstan S. C. Spencer. 1983.
"Employment-Output Conflicts, Factor-Price Distortions, and Choice of Tech-
nique: Empirical Results from Sierra Leone." *Economic Development and Cultural Change* 31:2 (January), 315–336.

Campos Icardo, Salvador. 1981. "Progress in Bilateral Relations." In Susan Kaufman Purcell, ed., *Mexico-United States Relations*. New York: Academy of Political Science.

Carlos, Manuel L. 1981. *State Policies, State Penetration, and Ecology: A Comparative Analysis of Uneven Development and Underdevelopment in Mexico's
Micro-Agrarian Regions*. Working Paper No. 19. La Jolla: University of California, San Diego, Center for U.S.-Mexican Studies.

Carr, Barry. 1983. "The Mexican Economic Debacle and the Labor Movement: A
New Era or More of the Same?" In Donald Wyman, ed., *Mexico's Economic
Crisis: Challenges and Opportunities*. Monograph No. 12. La Jolla: University of
California, San Diego, Center for U.S.-Mexican Studies.

CEICADAR. 1984. *CEICADAR: Aspectos generales de una experiencia en el
desarrollo agrícola regional*. Puebla, Mexico: CEICADAR.

Center for U.S.-Mexican Studies. 1984. "Program Agenda for 1984–1987." La
Jolla: University of California, San Diego, Center for U.S.-Mexican Studies.

CEPAL (Comisión Económica para América Latina). 1983. *Economía campesina y
agricultura empresarial*. México: Siglo Veintiuno.

Cernea, Michael. 1979. *Measuring Project Impact: Monitoring and Evaluation in
the PIDER Rural Development Project—Mexico*. Staff Working Paper No. 332.
Washington, D.C.: World Bank.

———. 1983. *A Social Methodology for Community Participation in Local Investments: The Experience of Mexico's PIDER Program*. Staff Working Paper No.
598. Washington, D.C.: World Bank.

Chuta, Enyinna, and Carl Liedholm. 1979. *Rural Non-Farm Employment: A Review of the State of the Art*. Rural Development Paper No. 4. East Lansing:
Michigan State University.

——. 1984. "Rural Small-Scale Industry: Empirical Evidence and Policy Issues." In Carl K. Eicher and John M. Staatz, eds., *Agricultural Development in the Third World*. Baltimore: Johns Hopkins University Press.

Cliffe, Lionel. 1978. "Labour Migration and Peasant Differentiation: Zambian Experiences." *Journal of Peasant Studies* 5, pp. 326–346.

Cohen, John M., and Norman Uphoff. 1977. *Rural Development Participation: Concepts and Measures for Project Design, Implementation, and Evaluation*. Rural Development Monograph No. 2. Ithaca, N.Y.: Cornell University, Center for International Studies.

Conroy, Michael E., Mario Coria Salas, and Felipe Vila Gonzáles. 1979. *Socio-Economic Incentives for Migration from Mexico to the U.S.: Magnitude, Recent Changes and Policy Implications*. Mexico-U.S. Migration Research Reports. Austin: University of Texas, Institute of Latin American Studies.

Cornelius, Wayne A. 1975. *Politics and the Migrant Poor in Mexico City*. Stanford, Calif.: Stanford University Press.

——. 1976a. "Mexican Migration to the United States: The View from Rural Sending Communities." Cambridge, Mass.: MIT, Center for International Studies.

——. 1976b. "Outmigration from Rural Mexican Communities." In *The Dynamics of Migration: International Migration*. Occasional Monograph Series No. 5, vol. 2. Washington, D.C.: Smithsonian Institution, Interdisciplinary Communications Program.

——. 1981a. *Mexican Migration to the United States: The Limits of Government Intervention*. Working Paper No. 5. La Jolla: University of California, San Diego, Center for U.S.-Mexican Studies.

——. 1981b. *The Future of Mexican Immigrants in California: A New Perspective for Public Policy*. Working Paper No. 6. La Jolla: University of California, San Diego, Center for U.S.-Mexican Studies.

——. 1981c. *Immigration, Mexican Development Policy, and the Future of U.S.-Mexican Relations*. Working Paper No. 8. La Jolla: University of California, San Diego, Center for U.S.-Mexican Studies.

——. 1983. "Simpson-Mazolli vs. the Realities of Mexican Immigration." In Wayne A. Cornelius and Ricardo Anzaldúa Montoya, eds., *America's New Immigration Law: Origins, Rationales, and Potential Consequences*. Monograph No. 11. La Jolla: University of California, Center for U.S.-Mexican Studies.

——. 1984. "Bitter Medicine, but Patient Survives." *Los Angeles Times*. September 5, 1984, p. 5.

——. 1985. "The Political Economy of Mexico under de la Madrid: Austerity, Routinized Crisis, and Nascent Recovery." *Estudios Mexicanos/Mexican Studies* 1:1 (Winter), 83–124.

——. 1986. *The Political Economy of Mexico under de la Madrid: The Crisis Deepens, 1985–1986*. Research Report No. 43. La Jolla: University of California, San Diego, Center for U.S.-Mexican Studies.

Cornelius, Wayne A., Leo R. Chavez, and Jorge A. Castro. 1982. *Mexican Immigrants and Southern California: A Summary of Current Knowledge*. Research

Report No. 36. La Jolla: University of California, San Diego, Center for U.S.-Mexican Studies.

Cornelius, Wayne A., and Ann L. Craig. 1984. *Politics in Mexico: An Introduction and Overview.* Reprint Series 1. La Jolla: University of California, San Diego, Center for U.S.-Mexican Studies.

Cornelius, Wayne A., and Ricardo Anzaldua Montoya, eds., 1983. *America's New Immigration Law: Origins, Rationales, and Potential Consequences.* Monograph No. 11. La Jolla: University of California, San Diego, Center for U.S.-Mexican Studies.

Corwin, Arthur, F., and Lawrence A. Cardoso. 1978. "Vamos al Norte: Causes of Mass Mexican Migration to the United States." In Arthur F. Corwin, ed., *Immigrants—and Immigrants: Perspectives on Mexican Labor Migration to the United States.* Westport, Conn.: Greenwood Press.

Craig, Ann L. 1981. *Mexican Immigration: Changing Terms of the Debate in the United States and Mexico.* Working Paper No. 4. La Jolla: University of California, San Diego, Center for U.S.-Mexican Studies.

———. 1983. *The First Agraristas: An Oral History of a Mexican Agrarian Reform Movement.* Berkeley: University of California Press.

Cross, Harry E., and James A. Sandos. 1981. *Across the Border: Rural Development in Mexico and Recent Migration to the United States.* Berkeley: University of California, Institute of Governmental Studies.

Dagodag, W. Tim. 1975. "Source Regions and Composition of Illegal Mexican Immigration to California." *International Migration Review* 9:4 (Winter), 499–511.

de la Madrid, Miguel. 1983. *Primer Informe de Gobierno, Sector Agropecuario Y Forestal.* Mexico City: Presidencia de la República.

de la Peña, Guillermo. 1981. *A Legacy of Promises: Agriculture, Politics, and Ritual in the Morelos Highlands of Mexico.* Austin: University of Texas Press.

del Castillo, Gustavo. 1985. *U.S.-Mexican Trade Relations: From the Generalized System of Preferences to a Formal Bilateral Trade Agreement.* Research Report No. 14. La Jolla: University of California, San Diego, Center for U.S.-Mexican Studies.

DeWalt, Billie R. 1979. *Modernization in a Mexican Ejido.* Cambridge: Cambridge University Press.

———. 1985. "Mexico's Second Green Revolution: Food for Feed." *Estudios Mexicanos/Mexican Studies* 1:1, 29–60.

Dinerman, Ina R. 1982. *Migrants and Stay-at-Homes: A Comparative Study of Rural Migration from Michoacán, Mexico.* Monograph No. 5. La Jolla: University of California, San Diego, Center for U.S.-Mexican Studies.

Domike, Arthur, and Louis Goodman. n.d. "Priorities for Agriculturally-Related Industries in Mexico." Manuscript, n.p.

Domínguez, Jorge I. 1982. Introduction to Jorge I. Domínguez, ed., *Mexico's Political Economy: Challenges at Home and Abroad.* Beverly Hills, Calif.: Sage Publications.

Ehrlich, Paul, Loy Belderback, and Anne Ehrlich. 1979. *The Golden Door: International Migration, Mexico, and the United States.* New York: Ballantine Books.

Eicher, Carl, and John Staatz, eds. 1984. *Agricultural Development in the Third World.* Baltimore, Md.: Johns Hopkins University Press.

Erasmus, Charles J. 1978. "Culture Change in Northwest Mexico." In Charles J. Erasmus, Solomon Miller, and Louis C. Faron, *Contemporary Change in Traditional Communities of Mexico and Peru.* Urbana: University of Illinois Press.

Erb, Guy F. 1982. *An American View of Mexican Trade Policy.* Working Paper No. 2. Washington, D.C.: Overseas Development Council.

Erb, Guy F., and Cathryn Thorup. 1984. *U.S.-Mexican Relations: The Issues Ahead.* Working Paper 35. Washington, D.C.: Overseas Development Council.

Erb, Richard D., and Stanley R. Ross, eds. 1981. *United States Relations with Mexico.* Washington, D.C.: American Enterprise Institute for Public Policy Research.

Esman, Milton J., and Norman T. Uphoff. 1984. *Local Organizations: Intermediaries in Rural Development.* Ithaca, N.Y.: Cornell University Press.

Fagen, Richard, and William S. Touhy. 1972. *Politics and Privilege in a Mexican City.* Stanford, Calif.: Stanford University Press.

Feder, Ernest. 1978. *Strawberry Imperialism.* Mexico City: Editorial Campesina.

Feinberg, Richard E. 1981. "Bureaucratic Organization and United States Policy toward Mexico." In Susan Kaufman Purcell, ed., *Mexico-United States Relations.* New York: Academy of Political Science.

Fernández, Raúl A. 1977. *The United-States-Mexico Border: A Politico-Economic Profile.* Notre Dame, Ind.: University of Notre Dame Press.

Fernández-Kelly, María Patricia. 1983a. *For We Are Sold, I and My People: Women and Industry in Mexico's Frontier.* Albany: State University of New York Press.

———. 1983b. "Alternative Education for Maquiladora Workers: Centro de Orientación de la Mujer Obrera." *Grassroots Development* 6:2 (Winter–Spring), 41–46.

Fonseca, Omar, and Lilia Moreno. n.d. *Jaripo.* Jilquilpan, Michoacán: Centro de Estudios de la Revolutión Mexicana Lázaro Cárdenas.

GAO (General Accounting Office). 1985. *Illegal Aliens: Information on Selected Countries' Employment Prohibition Laws.* Washington, D.C.: General Accounting Office.

Garrison, Helen. 1982. "Internal Migration in Mexico: A Test of the Todaro Model." *Food Research Institute Studies* 18:2, pp. 197–214.

Ghatak, Subrata, and Ken Ingersent. 1984. *Agriculture and Economic Development.* Baltimore, Md.: Johns Hopkins University Press.

Gil Días, Francisco. 1985. "Investment and Debt." In Peggy B. Musgrave, ed., *Mexico and the United States: Studies in Economic Interaction.* Boulder, Colo.: Westview Press.

Goldsmith, Arthur. 1985. "The Private Sector and Rural Development: Can Agribusiness Help the Small Farmer?" *World Development* 13:10–11 (October–November), 1125–1138.

Gotsch, Carl. 1974. "Economics, Institutions, and Employment Generation in Rural Areas." In Edgar O. Edwards, ed., *Employment in Developing Nations.* New York: Ford Foundation.

Gran, Guy. 1983. *Development by People: Citizen Construction of a Just World.* New York: Praeger.

Gregory, Peter. 1986. *The Myth of Market Failure: Employment and the Labor Market in Mexico.* Baltimore, Md.: Johns Hopkins University Press.

Griffiths, Stephen. 1979. "Migration and Enterpreneurship in a Philippine Peasant Village." *Papers in Anthropology* 20:1, pp. 127–144.

Grindle, Merilee S. 1977a. *Bureaucrats, Peasants, and Politicians in Mexico: A Case Study in Public Policy.* Berkeley: University of California Press.

———. 1977b. "Policy Change in an Authoritarian Regime: Mexico under Echeverría." *Journal of Inter American Studies and World Affairs* 19:4 (November), 523–555.

———. 1980. "The Implementor: Political Constraints on Rural Development in Mexico." In Merilee S. Grindle, ed., *Politics and Policy Implementation in the Third World.* Princeton, N.J.: Princeton University Press.

———. 1981. *Official Interpretations of Rural Underdevelopment: Mexico in the 1970's.* Working Paper No. 20. La Jolla: University of California, San Diego, Center for U.S.-Mexican Studies.

———. 1982. "Prospects for Integrated Rural Development: Evidence from Mexico and Colombia," *Studies in Comparative International Development* 27:3–4 (Fall–Winter), 124–149.

———. 1985. "Rhetoric, Reality, and Self-Sufficiency: Recent Initiatives in Mexican Rural Development." *Journal of Developing Areas* 19:2 (January), 171–184.

———. 1986. *State and Countryside: Development Policy and Agrarian Politics in Latin America.* Baltimore, Md.: Johns Hopkins University Press.

Haggblade, Steve, Carl Liedholm, and Donald C. Mead. 1986. *The Effect of Policy and Policy Reforms on Non-Agricultural Enterprises and Employment in Developing Countries: A Review of Past Experiences.* EEPA Discussion Paper No. 1. Cambridge, Mass.: Harvard Institute for International Development.

Hansen, Nils. 1985. "The Nature and Significance of Border Development Patterns." In Jay James Gibson and Alfonso Corona Renteria, eds., *The U.S. and Mexico: Borderland Development and the National Economies.* Boulder, Colo.: Westview Press.

Hansen, Roger. 1971. *The Politics of Mexican Development.* Baltimore, Md.: Johns Hopkins University Press.

———. 1979. *Beyond the North-South Stalemate.* New York: McGraw-Hill.

Harley, Richard. 1984. "The Greening of the Green Revolution Scientists: The Experience of Plan Puebla." Paper presented to the HIID Development Seminar, Cambridge, Mass., October 29.

Harris, John R., and Michael P. Todaro. 1970. "Migration, Unemployment and Development: A Two-Sector Analysis." *American Economic Review* 60:1 (March), 126–142.

Hart, Keith. 1982. *The Political Economy of West African Agriculture.* Cambridge: Cambridge University Press.

Hellman, Judith Adler. 1978. *Mexico in Crisis.* New York: Holmes and Meier.

———. 1983. "The Role of Ideology in Peasant Politics: Peasant Mobilization and Demobilization in the Laguna Region." *Journal of Inter-American Studies and World Affairs* 25:1, pp. 3–30.

Hewitt de Alcántara, Cynthia. 1976. *Modernizing Mexican Agriculture: Socioeconomic Implications of Technological Change, 1940–1970*. Geneva: United Nations Research Institute for Social Development.

Hufbauer, Gary Clyde, W. N. Harrell Smith IV, and Frank G. Vukmanic. 1981. "Bilateral Trade Relations." In Susan Kaufman Purcell, ed., *Mexico-United States Relations*. New York: Academy of Political Science.

Hugo, Graeme J. 1977. *Population Mobility in West Java*. Yogyakarta: Gadjah Mada University Press.

IDB (Interamerican Development Bank). Annual. *Economic and Social Progress in Latin America*. Washington, D.C.: IDB.

IMF (International Monetary Fund). 1985. *International Financial Statistics Yearbook*. Washington, D.C.: International Monetary Fund.

IMISAC. 1982. *Trabajadores de Michoacán: Historia de un pueblo migrante*. Morelia, Mexico: IMISAC.

International Migration Review. 1979. Special Issue on International Migration in Latin America 13 (Fall).

Jacobs, Michael. 1983. "Desarrollo regional y oferta de mano de obra: Campesinos y la industria de construcción en Tabasco 2000." Mimeographed. Xochimilco: Universidad Autónoma de Mexico.

Johnston, Bruce F., and William C. Clark. 1982. *Redesigning Rural Development: A Strategic Perspective*. Baltimore, Md.: Johns Hopkins University Press.

Johnston, Bruce, and Peter Kilby. 1975. *Agriculture and Structural Transformation: Economic Strategies in Late-Developing Countries*. New York: Oxford University Press.

Kearney, Michael. 1984. "Mixtec Migration from Oaxaca to Northwest Mexico and California." Paper prepared for presentation at the conference "Regional Aspects of U.S.-Mexican Integration: Past, Present, and Future." University of California, San Diego, Center for U.S.-Mexican Studies, May 21–22.

Kirk, Dudley. 1983. "Recent Demographic Trends and Present Population Prospects for Mexico." *Food Research Institute Studies* 19:1, pp. 93–111.

Kislev, Yoav, and Willis Peterson. 1986. "Economies of Scale in Agriculture: A Survey of the Evidence." Discussion Paper No. DRD 203. Washington, D.C.: World Bank, Development Research Department.

Laite, Julian. 1981. *Industrial Development and Migrant Labour in Latin America*. Austin: University of Texas Press.

LAM (*Latin American Monitor*). 1986. *Mexico 1986: Annual Report on Government, Economy, and Business*. London: Latin American Monitor.

Lassen, Cheryl A. 1980. *Landlessness and Rural Poverty in Latin America: Conditions, Trends, and Policies Affecting Income and Employment*. Special Series on Landlessness and Near-Landlessness. Ithaca, N.Y.: Cornell University, Center for International Studies.

Lele, Uma. 1984. "Rural Africa: Modernization, Equity, and Long-Term Development." In Carl K. Eicher and John M. Staatz, eds., *Agricultural Development in the Third World*. Baltimore, Md.: Johns Hopkins University Press.

Lele, Uma, and Arthur Goldsmith. 1986. "Building Agricultural Research Capacity: India's Experience with the Rockefeller Foundation and Its Significance for

Africa." Discussion Paper No. DRD213. Washington, D.C.: World Bank, Development Research Department.

Leonard, David. 1982. "Analyzing the Organizational Requirements for Serving the Rural Poor." In David K. Leonard and Dale Rogers Marshall, eds., *Institutions of Rural Development for the Poor: Decentralization and Organizational Linkages.* Berkeley: University of California, Institute of International Studies.

Levy, Daniel, and Gabriel Székely. 1983. *Mexico: Paradoxes of Stability and Change.* Boulder, Colo.: Westview Press.

Lewis, Oscar. 1951. *Life in a Mexican Village: Tepoztlan Restudied.* Urbana: University of Illinois Press.

Lewis, W. Arthur. 1954. "Economic Development with Unlimited Supplies of Labor." *Manchester School of Economic and Social Studies* 22, pp. 139–192.

Livingstone, Ian. 1986a. "Alternative Strategies for Development South of the Sahara: Contrasts between the ILO/JASPA and the World Bank Approaches." In Jobs and Skills Program for Africa of the International Labour Organization, *The Challenge of Employment and Basic Needs in Africa.* Nairobi: Oxford University Press.

——. 1986b. "The 'Sponge Effect': Population, Employment, and Incomes in Kenya." In Jobs and Skills Program for Africa of the International Labour Organization, *The Challenge of Employment and Basic Needs in Africa.* Nairobi: Oxford University Press.

Lomnitz, Claudio. 1978. "Cambios en la estructura de poder en Tepoztlán, 1920–1978." Thesis, Universidad Autónoma Mexicana-Iztapalapa, México.

Luiselli, Casio. 1982. *The Sistema Alimentario Mexicano (SAM): Elements of a Program of Accelerated Production of Basic Foodstuffs in Mexico.* Research Report 22. La Jolla: University of California, San Diego, Center for U.S.-Mexican Studies.

Malina, Robert M. 1980. *Push Factors in Mexican Migration to the United States: The Background to Migration: A Summary of Three Studies with Policy Implications.* Mexico-U.S. Migration Research Reports. Austin: University of Texas, Institute of Latin American Studies.

Mallon, Florencia. 1983. *The Defense of Community in Peru's Central Highlands: Peasant Struggle and Capitalist Transition, 1860–1940.* Princeton, N.J.: Princeton University Press.

Maram, Sheldon L. 1980. *Hispanic Workers in the Garment and Restaurant Industries in Los Angeles County: A Social and Economic Profile.* Working Paper No. 12. La Jolla: University of California, San Diego, Center for U.S.-Mexican Studies.

Mares, David R. 1982. "Agricultural Trade: Domestic Interests and Transnational Relations." In Jorge I. Dominguez, ed., *Mexico's Political Economy: Challenges at Home and Abroad.* Beverly Hills, Calif.: Sage Publications.

——. 1983. "Prospects for Mexico-U.S. Trade Relations in an Era of Economic Crisis and Restructuring." In Donald L. Wyman, ed., *Mexico's Economic Crisis: Challenges and Opportunities.* Monograph No. 12. La Jolla: University of California, San Diego, Center for U.S.-Mexican Studies.

Martínez Manautou, Jorge, ed., 1982. *The Demographic Revolution in Mexico.* Mexico: IMSS.

Massey, Douglas S., and Kathleen M. Schnabel. 1983. "Recent Trends in Hispanic Immigration to the United States." *International Migration Review* 17:2 (Summer), 212–244.

Mayagoitia, Alberto. 1976. *A Layman's Guide to Mexican Law*. Albuquerque: University of New Mexico Press.

Mead, Donald. 1984. "Of Contracts and Subcontracts: Small Firms in Vertically Dis-Integrated Production/Distribution Systems in LDC's." *World Development* 12:11–12 (November–December), 1095–1106.

Meissner, Frank. 1981. "The Mexican Food System (SAM: A Strategy for Sowing Petroleum." *Food Policy* 6:4 (November), 219–230.

Mellor, John W. 1976. *The New Economics of Growth: A Strategy for India and the Developing World*. Ithaca, N.Y.: Cornell University Press.

———. 1986. "Agriculture on the Road to Industrialization." In John P. Lewis and Valeriana Kallab, eds., *Development Strategies Reconsidered*. New Brunswick, N.J.: Transaction Books.

Mellor, John W., and Bruce F. Johnston. 1984. "The World Food Equation: Interrelations among Development, Employment, and Food Consumption." *Journal of Economic Literature* 22 (June), 531–574.

Mexico, Poder Ejecutivo Federal. 1983. *Programa Nacional de Alimentación, 1983–1988*. Mexico City: PEF.

———. 1985. *Programa Nacional de Desarrollo Rural Integral, 1985–1988*. Mexico City: PEF.

Meyer, Jean A. 1976. *The Cristero Rebellion: The Mexican People between Church and State, 1926–1929*. Cambridge: Cambridge University Press.

Meyer, Richard L. 1983. "Financing Rural Nonfarm Enterprises in Low-Income Countries." In J. D. Von Pischke, Dale W. Adams, and Gordon Donald, eds., *Rural Financial Markets in Developing Countries*. Baltimore, Md.: Johns Hopkins University Press.

———. 1985. "Rural Deposit Mobilization: An Alternative Approach for Developing Rural Financial Markets." Paper presented at the AID/IFAD Experts' Meeting on Small Farmer Credit, Rome, June 26–28.

Miller, Marjorie. 1984. "Factory Jobs Go Begging in Mexican Border Firms." *Los Angeles Times* (San Diego County), Sunday, April 29.

Mines, Richard. 1981. *Developing a Community Tradition of Migration: A Field Study in Rural Zacatecas, Mexico, and California Settlement Areas*. Monograph No. 3. La Jolla: University of California, San Diego, Center for U.S.-Mexican Studies.

Mines, Richard, and Douglas S. Massey. 1985. "Patterns of Migration to the United States from Two Mexican Communities." *Latin American Research Review* 20:2, pp. 104–123.

Montañez, Carlos, and Horacio Aburto. 1979. *Maíz: Política institucional y crisis agrícola*. Mexico City: Editorial Nueva Imagen.

Moris, Jon R. 1981. *Managing Induced Rural Development*. Bloomington, Ind.: International Development Institute.

Mukhoti, Bela. 1985. *Agriculture and Employment in Developing Countries: Strategies for Effective Rural Development*. Boulder, Colo.: Westview Press.

Murillo Castaño, Gabriel. 1984. *Migrant Workers in the Americas: A Comparative Study of Migration between Colombia and Venezuela and between Mexico and the United States.* Monograph No. 13. La Jolla: University of California, San Diego, Center for U.S.-Mexican Studies.

Musgrave, Peggy B., ed. 1985. *Mexico and the United States: Studies in Economic Interaction.* Boulder, Colo.: Westview Press.

NJ. Various issues. *National Journal.*

NYT. Various issues. *New York Times.*

ODC (Overseas Development Council). 1984a. "U.S. 'Costs' of Third World Recession: They Lose, We Lose." *Policy Focus,* no. 2.

———. 1984b. "The Debt Crisis: Can the Costs by Cut?" *Policy Focus,* no. 5.

———. 1984c. "The U.S. GSP Program: Trade Preferences and Development." *Policy Focus,* no. 6.

———. 1985. "U.S. Textiles and Apparel—Is Protection a Solution?" *Policy Focus,* no. 6.

Oficina de Asesores del C. Presidente. 1980. "SAM: Primer planteamiento de metas de consumo y estrategia de producción de alimentos básicos para 1980–1982." México City: OAP.

Owen, Roger. 1985. *Migrant Workers in the Gulf.* Report No. 68. London: Minority Rights Group.

Page, John M., Jr., and William F. Steel. 1984. *Small Enterprise Development: Economic Issues from African Experience.* Technical Paper 26. Washington, D.C.: World Bank.

Passel, Jeffrey S., and Karen A. Woodrow. 1984. "Geographic Distribution of Undocumented Immigrants: Estimates of Undocumented Aliens Counted in the 1980 Census by State." Paper presented at the 1984 meeting of the Population Association of America, Minneapolis, Minnesota, May 3–5.

Pellicer de Brody, Olga. 1981. "A Mexican Perspective." In Susan Kaufman Purcell, ed., *Mexico-United States Relations.* New York: Academy of Political Science.

Piore, Michael J. 1979. *Birds of Passage: Migrant Labor and Industrial Societies.* Cambridge: Cambridge University Press.

Portes, Alejandro. 1982. "Of Borders and States: A Skeptical Note on the Legislative Control of Immigration." Paper presented at the symposium "America's New Immigration Law," University of California, San Diego, November 19–20.

Puchala, Donald J., and Jane Staveley. 1979. "The Political Economy of Taiwanese Agricultural Development." In Raymond F. Hopkins, Donald J. Puchala, and Ross B. Talbot, eds., *Food, Politics, and Agricultural Development: Case Studies in the Public Policy of Rural Modernization.* Boulder, Colo.: Westview Press.

Purcell, John F. H. 1982. *Trade Conflicts and U.S.-Mexican Relations.* Working Paper No. 38. La Jolla: University of California, San Diego, Center for U.S.-Mexican Studies.

Purcell, Susan Kaufman. 1975. *The Mexican Profit-Sharing Decision: Politics in an Authoritarian Regime.* Berkeley: University of California Press.

———. 1981. "Business-Government Relations in Mexico: The Case of the Sugar Industry." *Comparative Politics* 13:2 (January), 211–233.

Purcell, Susan, and John Purcell. 1980. "State and Society in Mexico." *World Politics* 32, pp. 194–227.

Redclift, Michael. n.d. "Production Programs for Small Farmers: Plan Puebla as Myth and Reality." Mexico City: Ford Foundation.

———. 1981. *Development Policymaking in Mexico: The Sistema Alimentario Mexicano.* Working Paper No. 4. La Jolla: University of California, San Diego, Center for U.S.-Mexican Studies.

Redfield, Robert. 1930. *Tepoztlán—A Mexican Village.* Chicago: University of Chicago Press.

Reich, Michael R., Yasuo Endo, and C. Peter Timmer. 1986. "Agriculture: The Political Economy of Structural Change." In Thomas K. McCraw, ed., *America versus Japan.* Boston: Harvard Business School Press.

Reichert, Joshua S. 1981. "The Migrant Syndrome: Seasonal U.S. Wage Labor and Rural Development in Central Mexico." *Human Organization* 40:1 (Spring), 56–66.

Reichert, Joshua S., and Douglas S. Massey. 1980. "History and Trends in U.S. Bound Migration from a Mexican Town." *International Migration Review* 14:4 (Winter), 475–491.

Reilly, Charles. 1983. "The Arizona Farmworkers Union and Mexican Rural Development." Center for U.S.-Mexican Studies, University of California, San Diego.

Reisler, Mark. 1976. *By the Sweat of Their Brow: Mexican Immigrant Labor in the United States, 1900–1940.* Westport, Conn.: Greenwood Press.

Rempel, Henry, and Richard Lobdell. 1978. "The Role of Urban to Rural Remittances in Rural Development." *Journal of Development Studies* 14, pp. 324–341.

Reubens, Edwin P. 1979. "Surplus Labor, Emigration, and Public Policies: Requirements for Labor Absorption in Mexico." In Barry W. Poulson and T. Noel Osborn, eds., *U.S.-Mexico Economic Relations.* Boulder, Colo.: Westview Press, 1979.

Reyna, José Luís. 1974. *Control político: Estabilidad y desarrollo en México.* Cuadernos del CES 3. México: El Colegio de México.

Reynolds, Clark W. 1979. "Labor Market Projections for the United States and Mexico and Current Migration Controversies." *Food Research Institute Studies* 17:2, pp. 121–155.

———. 1983. "Mexico's Economic Crisis and the United States: Toward a Rational Response." In Donald L. Wyman, ed., *Mexico's Economic Crisis: Challenges and Opportunities.* Monograph No. 12. La Jolla: University of California, San Diego, Center for U.S.-Mexican Studies.

Rhoda, Richard. 1983. "Rural Development and Urban Migration: Can We Keep Them Down on the Farm?" *International Migration Review* 17 (Spring), pp. 34–64.

Rico, Carlos. 1981. "Prospects for Economic Cooperation." In Susan Kaufman Purcell, ed., *Mexico-United States Relations.* New York: Academy of Political Science.

Riding, Alan. 1985. *Distant Neighbors: A Portrait of the Mexicans.* New York: Alfred A. Knopf.

Rivière d'Arc, Helene. 1980. "Change and Rural Emigration in Central Mexico." In David A Preston, ed., *Environment, Society, and Rural Change in Latin America.* Chichester: John Wiley.

Roberts, Kenneth D. 1982. "Agrarian Structure and Labor Mobility in Rural Mexico," *Population and Development Review* 8:2 (June), 299–322.

——. 1984. "The Impact of U.S. Technology on the Mexican Bajio: Seeds, Sorghum, and Socioeconomic Change." Paper presented at the conference "Regional Aspects of U.S.-Mexican Integration," University of California, San Diego, Center for U.S.-Mexican Studies, May 21–22.

——. 1985. "Household Labour Mobility in a Modern Agrarian Economy: Mexico." In Guy Standing, ed., *Labour Circulation and the Labour Process.* London: Croom Helm.

Rodríguez Pina, Javier, and Martha Loyo Camacho. 1983. "El movimiento perpetuo: La migración reciente de trabajadores mexicanos a Estados Unidos, 1942–1982." *Azcapotzalco* 4:8 (January–April), 9–26.

Rondinelli, Dennis A., and Kenneth Ruddle. 1978. *Urbanization and Rural Development: A Spatial Policy for Equitable Growth.* New York: Praeger.

Ronfeldt, David. 1973. *Atencingo: The Politics of Agrarian Struggle in a Mexican Ejido.* Stanford, Calif.: Stanford University Press.

Ronfeldt, David, Richard Nehring, and Arturo Gándara. 1980. *Mexico's Petroleum and U.S. Policy: Implications for the 1980s.* Santa Monica, Calif.: Rand Corporation.

SALA. 1983. *Statistical Abstract for Latin America.* Los Angeles: University of California, Center for Latin American Studies.

Samet, Andrew James, and Gary Clyde Hufbauer. 1982. "'Unfair' Trade Practices: A Mexican-American Drama." Working Paper No. 1. Washington, D.C.: Overseas Development Council, U.S.-Mexico Project.

Sánchez, Guadelupe L., and Jesus Romo. 1981. *Organizing Mexican Undocumented Farm Workers on Both Sides of the Border.* Working Paper No. 27. La Jolla: University of California, San Diego, Center for U.S.-Mexican Studies.

Sánchez Ugarte, Fernando. 1983. "The Role of Subsidies and Taxes in Mexican Agricultural Policy: Taxation in Agriculture." Paper prepared for presentation at the workshop "U.S.-Mexico Agricultural Trade, Pricing Policy, and Resources," Santa Fe, New Mexico, August 22–25.

Sanderson, Steven E. 1981. *The Receding Frontier: Aspects of the Internationalization of U.S.-Mexican Agriculture and Their Implications for Bilateral Relations in the 1980s.* Working Paper No. 15. La Jolla: University of California, San Diego, Center for U.S. Mexican Studies.

——. 1983. "The Complex No-Policy Option: U.S. Agricultural Relations with Mexico." Paper prepared for the workshop on "Mexico-U.S. Agricultural Trade, Pricing Policy, and Resources," Santa Fe, New Mexico, August 22–25.

——. 1986. *The Transformation of Mexican Agriculture: International Structure and the Politics of Rural Change.* Princeton, N.J.: Princeton University Press.

Sanderson, Susan Walsh. 1984. *Land Reform in Mexico, 1910–1980.* Orlando, Fla.: Academic Press.

Schmink, Marianne. "Household Economic Strategies: Review and Research Agenda." *Latin American Research Review* 19:3, pp. 87–101.

Schryer, Frans J. 1980. *The Rancheros of Pisaflores: A History of a Peasant Bourgeoisie in Twentieth-Century Mexico.* Toronto: University of Toronto Press.

Schumacher, August. 1981. *Agricultural Development and Rural Employment: A Mexican Dilemma.* Working Paper No. 21. La Jolla: University of California, San Diego, Center for U.S.-Mexican Studies.

Scott, Christopher D. 1980. "Transnational Corporations and the Food Industry in Latin America: An Analysis of the Determinants of Investment and Divestment." Working Paper No. 64, Latin American Program, Woodrow Wilson Center, Washington, D.C.

Seligson, Mitchell, and Edward J. Williams. 1981. *Maquiladoras and Migration: Workers in the Mexican–United States Border Industrialization Program.* Austin: Texas Press Services.

Sewell, John W., Richard E. Feinberg, and Valeriana Kallab, eds. 1985. *U.S. Foreign Policy and the Third World: Agenda, 1985–86.* Washington, D.C.: Overseas Development Council.

Shaw, M. M., and F. Willekens. 1978. *Rural Urban Population Projections for Kenya and Implications for Development.* Laxenburg, Austria: IIASA.

Silvers, Arthur, and Pierre Crosson. 1980. *Rural Development and Urban-Bound Migration in Mexico.* Washington, D.C.: Resources for the Future.

Smith, Peter H. 1979. *Labyrinths of Power: Political Recruitment in Twentieth-Century Mexico.* Princeton, N.J.: Princeton University Press.

———. 1980. *Mexico: The Quest for a U.S. Policy.* New York: Foreign Policy Association.

———. 1984a. "Mexico: The Continuing Quest for a Policy." In Richard Newfarmer, ed., *From Gunboats to Diplomacy: New U.S. Policies for Latin America.* Baltimore, Md.: Johns Hopkins University Press.

———. 1984b. *Mexico, Neighbor in Transition.* New York: Foreign Policy Association.

———. 1985. "U.S.-Mexican Relations: The 1980s and Beyond." *Journal of Interamerican Studies and World Affairs* 27:1 (February), 91–101.

Snodgrass, Donald R. 1979. *Small-Scale Manufacturing Industries: Patterns, Trends, and Possible Policies.* Discussion Paper No. 54. Cambridge, Mass.: Harvard University, Harvard Institute for International Development.

Spalding, Rose J. 1984. *The Mexican Food Crisis: An Analysis of the SAM.* Research Report No. 33. La Jolla: University of California, San Diego, Center for U.S.-Mexican Studies.

———. 1985. "Structural Barriers to Food Programming: An Analysis of the 'Mexican Food System.'" *World Development* 13:12 (December), 1249–1262.

SPP (Secretaria de Programacion y Presupuesto). 1982. *Anuario estadístico de los Estados Unidos Mexicanos 1981.* Mexico City: SPP.

———. 1983a. *México: Información sobre aspectos geográficos, sociales, y económicos,* vol. 3. Mexico City, SPP.

———. 1983b. *Plan Nacional de Desarrollo, 1983–1988.* Mexico City: SPP.

Squire, Lyn. 1981. *Employment Policy in Developing Countries: A Survey of Issues and Evidence.* New York: Oxford University Press.

Staatz, John M., and Carl K. Eicher. 1984. "Agricultural Development Ideas in Historical Perspective." In Carl K. Eicher and John M. Staatz, eds., *Agricultural Development in the Third World*. Baltimore, Md.: Johns Hopkins University Press.

Stark, Oded, and D. Levhari. 1982. "On Migration and Risk in LDCs." *Economic Development and Cultural Change* 31:1, 191–196.

Stark, Oded, and J. Edward Taylor. 1985. *Remittances and Inequality*. Discussion Paper No. 16. Cambridge, Mass.: Harvard University, Migration and Development Program.

Stevens, Evelyn P. 1974. *Protest and Response in Mexico*. Cambridge, Mass.: MIT Press.

Story, Dale. 1982. "Trade Politics in the Third World: A Case Study of the Mexican GATT Decision." *International Organization* 36:4 (Autumn), 767–794.

Stuart, James, and Michael Kearney. 1981. *Causes and Effects of Agricultural Labor Migration from the Mixteca of Oaxaca to California*. Working Paper No. 28. La Jolla: University of California, San Diego, Center for U.S.-Mexican Studies.

Taylor, J. Edward. 1984a. *Migration Networks and Risk in Household Labor Decisions: A Study of Migration from Two Mexican Villages*. Ph.D. diss., University of California, Berkeley.

———. 1984b. *Differential Migration, Networks, Information, and Risk*. Discussion Paper No. 11. Cambridge, Mass.: Harvard University, Migration and Development Program.

Teitelbaum, Michael S. 1985. *Latin Migration North: The Problem for U.S. Foreign Policy*. New York: Council on Foreign Relations.

Tejera Gaona, Hector. 1982. *Capitalismo y campesinado en el Bajío: Un estudio de caso*. Mexico: Instituto National de Antropología e Historia.

Tendler, Judith. 1982. *Turning Private Voluntary Organizations into Development Agencies: Questions for Evaluation*. AID Program Evaluation Discussion Paper No. 12. Washington, D.C.: U.S. Agency for International Development.

Thiesenhusen, William C. 1987. "Rural Development Questions in Latin America." *Latin American Research Review* 22:1, pp. 171–203.

Thomas, John Woodward. 1974. "Employment Creating Public Works Programs: Observations on Political and Social Dimensions." In Edgar O. Edwards, ed., *Employment in Developing Nations*. New York: Ford Foundation.

Thompson, Gary, Ricardo Amon, and Philip L. Martin. 1985. "Agricultural Development and Emigration: Rhetoric and Reality." University of California, Davis, Department of Agricultural Economics.

Thorup, Cathryn L. 1986. *The United States and Mexico: Face to Face with New Technology*. Washington, D.C.: Overseas Development Council.

Timmer, C. Peter. 1986. "Review Article: *Redesigning Rural Development* from a Food Policy Perspective." *Economic Development and Cultural Change* 34:4 (July), 855–860.

Timmer, C. Peter, Walter P. Falcon, and Scott R. Pearson. 1983. *Food Policy Analysis*. Baltimore, Md.: Johns Hopkins University Press.

Todaro, Michael P. 1969. "A Model of Labor Migration and Urban Unemployment in Less Developed Countries." *American Economic Review* 59:1 (March), 138–148.

Trebat, Thomas J. 1985. "Mexico's Foreign Financing." In Peggy B. Musgrave, ed., *Mexico and the United States: Studies in Economic Interaction.* Boulder, Colo.: Westview Press.

Universidad Autónoma de Baja California, Instituto de Investigaciones Sociales. 1983. *Migración y absorción de mano de obra en los asentamientos humanos irregulares de la ciudad de Mexicali, B.C.: 1940–1982.* Reporte Terminal del Programa de Investigación. Mexicali: Universidad Autónoma de Baja California.

Vaitsos, Constantine V. 1974. "Employment Effects of Foreign Direct Investments in Developing Countries." In Edgar O. Edwards, ed., *Employment in Developing Nations.* New York: Ford Foundation.

Valdes, Alberto. 1986. "Latin American Regional Paper." Paper Prepared for the Study on U.S. Agricultural Exports and Third World Development. N.p.: Curry Foundation.

Verduzco I., Gustavo. 1984. "Una ciudad agrícola: Trajectoría de la agricultura zamorana." Paper prepared for presentation at a conference on "Regional Aspects of U.S.-Mexican Integration: Past, Present, and Future." University of California, San Diego, Center for U.S.-Mexican Studies, May 21–22.

Von Pischke, J. D., Dale W. Adams, and Gordon Donald. 1983. *Rural Financial Markets in Developing Countries: Their Use and Abuse.* Baltimore, Md.: Johns Hopkins University Press.

Ward, Peter. 1986. *Welfare Politics in Mexico: Papering over the Cracks.* London: Allen and Unwin.

Warman, Arturo. 1980. *"We Come to Object": The Peasants of Morelos and the National State.* Baltimore, Md.: Johns Hopkins University Press.

Weinert, Richard S. 1983. "International Finance: Banks and Bankruptcy." *Foreign Policy* 50 (Spring), 138–149.

Weintraub, Sidney. 1984. *Free Trade between Mexico and the United States?* Washington, D.C.: Brookings Institution.

———. 1985. "Trade and Structural Change." In Peggy B. Musgrave, ed., *Mexico and the United States: Studies in Economic Interaction.* Boulder, Colo.: Westview Press.

Whiteford, Scott. n.d. "Linkage, Process, and Structure: The Mexicali Case."

Whiting, Van R., Jr. 1983. "Markets and Bargains: Foreign Investment and Development Strategies in Mexico." In Donald L. Wyman, ed., *Mexico's Economic Crisis: Challenges and Opportunities.* Monograph No. 12. La Jolla: University of California, San Diego, Center for U.S.-Mexican Studies.

———. 1984. *The Politics of Technology Transfer in Mexico.* Research Report No. 37. La Jolla: University of California, San Diego, Center for U.S.-Mexican Studies.

Whyte, William F. 1981. *Participatory Approaches to Agricultural Research and Development.* Ithaca, N.Y.: Cornell University, Rural Development Committee.

Winnie, Jr., William W., Elsa Guzman Flores, and Victor M. Hernandez-Saldana. 1979. "Migration from West Mexico to the United States." In Barry W. Poulson and T. Noel Osborn, eds., *U.S.-Mexico Economic Relations.* Boulder, Colo.: Westview Press, 1979.

Wolf, Eric. 1959. *Sons of the Shaking Earth*. Chicago: University of Chicago Press.
Womack, John, Jr. 1968. *Zapata and the Mexican Revolution*. New York: Vintage Books.
World Bank. 1978. *Rural Enterprise and Nonfarm Employment*. Washington, D.C.: WB.
——. 1984. *World Development Report*. Washington, D.C.: WB.
——. 1985. *World Development Report*. Washington, D.C.: WB.
——. 1986. *World Development Report*. Washington, D.C.: WB.
——. 1987. *World Development Report*. Washington, D.C.: WB.
WSJ. Various issues. *Wall Street Journal*.
Wyman, Donald L. 1981. *The United States Congress and the Making of U.S. Policy toward Mexico*. Working Paper No. 13. La Jolla: University of California, San Diego, Center for U.S.-Mexican Studies.
Wyman, Donald L., ed. 1983. *Mexico's Economic Crisis: Challenges and Opportunities*. Monograph No. 12. La Jolla: University of California at San Diego, Center for U.S.-Mexican Studies.
Yap, L. 1977. "The Attraction of Cities: A Review of the Migration Literature." *Journal of Development Economics* 4, pp. 239–264.
Yates, P. Lamartine. 1981. *Mexico's Agricultural Dilemma*. Tucson: University of Arizona Press.
Zazueta, Carlos H., and Manuel García y Griego. 1982. *Los trabajadores mexicanos en Estados Unidos: Resultados de la encuesta nacional de emigración a la frontera norte del país y a los Estados Unidos*. Mexico: Centro Nacional de Información y Estadísticas del Trabajo, Secretaría del Trabajo y Provisión Social.
Zolberg, Aristide R. 1982. "Dilemmas at the Gate: The Politics of Immigration in Advanced Industrial Societies." Paper presented at the symposium "America's New Immigration Law," University of California, San Diego, Center for U.S.-Mexican Studies, November 19–20.

Index

Library of Congress Cataloging-in-Publication Data

Grindle, Merilee Serrill.
  Searching for rural development.

  (Food systems and agrarian change)
  Bibliography: p.
  Includes index.
  1. Migrant labor—Mexico. 2. Labor supply—Mexico. 3. Rural development—
Mexico. I. Title. II. Series.
HD5856.M6G74 1988   331.6'2'0972   87-47970
ISBN 0-8014-2109-8 (alk. paper)